Extension: From Brush to Algorithm

SANDEEP CHAVAN

Published by SANDEEP CHAVAN, 2024.

EXTENSION: FROM BRUSH TO ALGORITHM

First edition. December 15, 2024.

ISBN: 979-8230658825

Written by SANDEEP CHAVAN.

Table of Contents

Introduction 1
1. The Timeless Quest for Creativity........ 3
Part I: Creativity and Tools – A Historical Perspective 23
2: From Brush to Algorithm 25
3: The Machine Age 51
Part II: The Paradox of Technology in Creativity 79
4: A New Frontier........ 81
5: Creativity without Tools 107
6: The Dilemma of Modern Creators........ 135
Part III: Restrictions and Their Consequences........ 177
7: The Fear of Machines 179
8. Universities and Old Lenses........ 205
Part IV: Redefining Creativity for the Future 223
9: The Essence of Creativity 225
10: Humans and Machines........ 249
Part V: Solutions and Actionable Frameworks 273
11: Rethinking Policies 275
12: Balanced Creative Identity 293
13: The Universal Skill of Creativity 303
Conclusion........ 311
14: The Evolving Spirit of Creativity........ 313
Appendices........ 319
Acknowledgment 323
About the Author........ 325

To all the creators—The visionaries, the dreamers, the innovators, and the storytellers.

This book is for those who see tools not as replacements but as extensions of their limitless imagination.

May you continue to shape the world with your creativity, weaving humanity into every brushstroke, every note, every word, and every algorithm.

Your spirit defines the future.

"The tools we shape, in turn, shape us—but it is the human spirit that breathes life into creation."— Anonymous

Disclaimer

The information, ideas, and concepts presented in *Extension: From Brush to Algorithm – The Human Spirit with or Without Tools* are intended for educational and informational purposes only. While every effort has been made to ensure accuracy, the author and publisher make no representations or warranties regarding the completeness, reliability, or applicability of the content to any specific situation.

This book explores the evolving intersection of creativity and technology, offering examples, strategies, and opinions that reflect the author's perspectives. Readers are encouraged to thoughtfully adapt the insights and approaches discussed to their unique circumstances, while remaining mindful of the ethical and practical considerations when integrating tools like AI into their personal or professional practices.

The content within this book does not constitute technical, legal, or professional advice. Readers seeking guidance tailored to their field or industry are advised to consult qualified professionals or authoritative sources. Any reliance on the information provided is undertaken strictly at the reader's own discretion and risk.

The author and publisher disclaim any liability for outcomes or consequences, direct or indirect, arising from the use or application of the material presented in this book. References to specific tools, platforms, technologies, or systems are made in good faith for illustrative purposes only and do not constitute endorsements or recommendations of those products or services.

This book is a work of intellectual and creative exploration. Any similarities to real individuals, organizations, or events are purely coincidental unless explicitly stated. The opinions expressed are solely those of the author and do not necessarily reflect the views of any affiliated organizations, institutions, or entities.

Readers are encouraged to approach the concepts and ideas in this book with an open mind, critical thinking, and a strong commitment to ethical and responsible practices. Creativity, when paired with thoughtful intent, remains humanity's most powerful force for innovation and progress.

Preface

Creativity has always been humanity's most powerful gift. It is the force that drives invention, shapes art, solves complex problems, and inspires progress. From the earliest cave paintings etched with bare hands to intricate algorithms creating breathtaking digital landscapes, the human spirit has always found ways to extend itself through tools. At each stage of civilization, tools have enhanced our ability to create—but they have never replaced the essence of creativity itself.

As we stand at the crossroads of a technological revolution, questions surrounding creativity and tools have gained new urgency. Will machines diminish human creativity, or will they elevate it? Are tools merely extensions of human ingenuity, or do they redefine its boundaries? *Extension: From Brush to Algorithm* seeks to explore these critical questions while celebrating the enduring role of the human spirit.

This book is not just about technology, nor is it about art in isolation. It is about the evolving relationship between the creator and their tools—how innovation reshapes the tools we use and, in turn, how those tools push us to evolve as creators. It highlights how each generation has seen its own set of tools emerge and transform fields, yet the timeless core of creativity remains distinctly human—anchored in our emotions, experiences, and aspirations.

The title itself, *Extension: From Brush to Algorithm*, reflects the journey of humanity's creative expression. Whether it's a brush delicately applying paint to canvas, a sculptor's chisel carving marble, or a machine-learning algorithm generating patterns from complex data, every tool we create is, at its core, an extension of our minds and hands. Tools are not replacements; they are amplifiers of our ideas, inspirations, and capabilities.

In this book, I attempt to navigate the purist's perspective, the modern creator's dilemma, and the role of institutions in shaping the future of creativity. I examine how tools like artificial intelligence have opened unimaginable opportunities but also sparked debates about ethics, authenticity, and the limits of technology. Ultimately, the book calls for a balanced approach—one that integrates the strengths of technology without compromising the soul of human creativity.

To every artist, designer, writer, scientist, and creator of any kind, this book is for you. It is a conversation about where we are and where we are headed as creators in the machine age. It is a reminder that the human spirit, with or without tools, is timeless, evolving, and resilient.

May this work inspire you to see tools—whether brushes, algorithms, or technologies yet to come—not as threats, but as companions in your creative journey. For creativity is not confined by tools; it is powered by the unrelenting spirit of those who dream, create, and reimagine the world around them.

Sandeep Chavan

Introduction

"Tools have always been the silent partners of human creativity—not defining what we create, but amplifying how we imagine. To understand the future of innovation, we must first look back at the timeless interplay between the mind and the tools it wields."

— Sandeep Chavan

1. The Timeless Quest for Creativity

Creativity is the lifeblood of humanity, an electrifying force that has ignited our journey from primitive survival to modern marvels. Imagine the wonder of ancient cave dwellers capturing their stories on stone walls, their crude tools scratching the first strokes of artistic expression. Fast forward to today's digital age, where creativity shapes breathtaking virtual landscapes and pushes the boundaries of human potential. At its core, creativity is more than just an idea or a fleeting spark—it's a bold, unstoppable drive that fuels progress, challenges conventions, and redefines possibilities.

Throughout history, creativity has been our secret weapon, a survival instinct transformed into a dazzling medium for self-expression. It has danced through time, adapting and thriving amidst the challenges of every era, proving its resilience and timeless allure. As the tools we wield evolve—whether stone chisels or artificial intelligence—so too does the very essence of what it means to create. This chapter embarks on an exhilarating exploration of creativity's eternal journey, uncovering its shifting definitions across ages and the transformative power of the tools that amplify our limitless imagination.

"Tools may amplify our hands, but it is the boundless human imagination that turns them into instruments of wonder."

1.1 Defining Creativity across Ages

Creativity is the heartbeat of humanity, an ever-evolving force that has guided us from the dawn of survival to the heights of artistic expression. It is in the sparks of fire lit by ancient hands, in the stories etched on cave walls, and in the masterpieces of modern innovation. Across ages, creativity has been our compass, pointing not just to what is necessary but to what is possible. It is the thread that binds our imagination to reality, adapting to challenges, inspiring change, and carving pathways into the unknown. This journey of creativity is as timeless as it is transformative, revealing the essence of who we are and who we can become.

The Evolution of Creativity: From Survival to Expression

Creativity began as a necessity—a raw, instinctive force essential for survival. In the unforgiving landscapes of early human existence, it was the difference between life and death. Our ancestors, faced with predators, harsh climates, and scarcity of resources, developed tools by chipping stones, discovered fire to cook and ward off the cold, and built rudimentary shelters for safety. These acts of creation were not artistic endeavors but practical solutions born out of sheer necessity.

Yet, even in these primal acts, there was a flicker of something extraordinary. When early humans took charcoal and ochre to paint animals on cave walls or carved rudimentary figures from bone and ivory, they transcended the realm of mere survival. These expressions were not just utilitarian—they were symbolic, emotional, and deeply human. In these acts, creativity began to evolve into a medium of self-expression, a language that conveyed their fears, dreams, and understanding of the world around them.

This shift from survival to expression was transformative, marking a significant milestone in the human journey. Creativity became more than a tool for enduring the harshness of life; it became a testament to the human spirit's yearning to leave a mark. The cave paintings of Lascaux, the Venus figurines, and other early artistic endeavors were not merely decorative. They were statements—a way to connect with others, to tell stories, and to imagine realities beyond the immediate.

As humans continued to adapt and grow, this newfound dimension of creativity wove itself into the very fabric of human identity. The practical skills that once ensured survival evolved into intricate art forms. Humans began to see the world not just as it was but as it could be, shaping reality with the power of imagination. What started as a means to survive became a way to thrive, inspiring collaboration, innovation, and cultural development.

This evolution of creativity, from a raw instinct to an elevated expression, reflects the profound adaptability of the human mind. It highlights our capacity to transform challenges into opportunities and necessity into beauty. It reminds us that creativity is not just an act—it is a defining feature of what it means to be human, bridging the gap between survival and the aspiration for meaning and connection.

Creativity as a Reflection of Human Imagination

As humans evolved, so did the scope of their imagination. Creativity, once rooted in the necessity of survival, expanded into a profound reflection of what could be—not just what was. It became a mirror, capturing not only the immediate realities of human existence but also the vast possibilities of the future. Through creativity, humans found a way to express their dreams, fears, and aspirations, venturing beyond the constraints of logic and practicality. It was a force that allowed them to envision futures yet to unfold, create beauty from the mundane, and solve problems in ways that were as innovative as they were poetic.

The crafting of tools gave way to the crafting of ideas. A sharpened stone, once a simple means of hunting, became an artifact etched with intricate designs. A functional clay pot transformed into a canvas for artistry. Creativity began to transcend utility, becoming an outlet for the human spirit to express itself in profound ways. It turned the utilitarian into the poetic, the mundane into the extraordinary.

"The true sign of intelligence is not knowledge but imagination." – Albert Einstein.

This insight beautifully encapsulates the transformative power of human imagination. Imagination gave rise to innovation, enabling humans to bridge the gap between the tangible and the intangible. It fueled the myths of ancient storytellers, weaving narratives that connected generations and preserved wisdom. These stories were more than entertainment—they were frameworks for understanding the world and imagining what lay beyond.

Creativity also found expression in the architectural marvels of civilizations. The soaring pyramids of Egypt, the intricate carvings of Angkor Wat, and the majestic Parthenon in Greece are all testaments to how imagination shapes the tangible. These creations were not merely functional; they were symbolic, reflecting the aspirations, beliefs, and ingenuity of the societies that built them.

Through imagination, creativity transforms thoughts into reality and connects the abstract to the concrete. It allows the human mind to explore boundless potential, turning fleeting ideas into enduring legacies. As author Maya Angelou once said, *"You can't use up creativity. The more you use, the more you have."* This boundless nature of creativity reflects the infinite possibilities that lie within the human spirit, waiting to be imagined and brought to life.

Creativity as a reflection of human imagination is a celebration of humanity's greatest gift—the ability to see beyond the immediate, to innovate, and to dream. It is this unyielding force that has propelled us forward, bridging the gap between what is and what could be, transforming not only the world around us but also the way we perceive it.

Cultural Perspectives on Creativity Through History

Creativity, though universal, finds its unique expression in every culture, reflecting the values, traditions, and aspirations of its people. Across time and geography, it has served as both a personal pursuit and a collective endeavor, shaping identities and bridging generations. Each culture's relationship with creativity highlights its significance as a cornerstone of human existence, weaving together the spiritual, communal, and intellectual threads of society.

In ancient India, creativity was deeply intertwined with spirituality and storytelling. The grand temples, etched with intricate sculptures, were more than places of worship—they were canvases for devotion and narratives carved in stone. Every detail, from the delicate expressions of deities to the complex depictions of epics like the Ramayana and Mahabharata, told stories of faith, morality, and human nature. These creations were as much a testament to artistic ingenuity as they were to cultural and spiritual identity.

In Africa, the rhythmic patterns of drumming resonate with the heartbeat of community and tradition. These rhythms are not mere sounds but carriers of stories, histories, and emotions, passed down through generations. Each beat speaks of celebration, mourning, or resistance, demonstrating how creativity becomes a language that connects people to their roots while fostering a sense of unity.

In East Asia, the art of calligraphy stands as a vivid example of creativity blending beauty with meaning. Each brushstroke, precise and deliberate, transforms written language into a visual masterpiece. The flowing characters are not just symbols but expressions of harmony, discipline, and cultural pride. Calligraphy in countries like China and Japan represents an art form where the act of writing becomes a meditative practice, merging aesthetics with the communication of ideas.

Europe's Renaissance, often regarded as a golden age of creativity, saw the collision of science and art to produce works that have endured through centuries. Figures like Leonardo da Vinci and Michelangelo exemplify how creativity transcended boundaries, merging disciplines to redefine human understanding and artistic expression. The Renaissance was more than a cultural movement; it was a reawakening of the human spirit, driven by the belief in limitless possibilities.

In other parts of the world, creativity took on roles that balanced spiritual, communal, and personal expression. The indigenous art of Australia, with its vivid depictions of the Dreamtime, conveys profound spiritual beliefs while preserving ancestral knowledge. The vibrant textiles of South America tell stories of heritage and identity, reflecting the creativity of cultures that value harmony with nature.

These cultural imprints remind us that creativity is both universal and deeply personal. It adapts to the values, needs, and aspirations of each era, becoming a lens through which humanity views and shapes its existence.

As Pablo Picasso once said,

"The purpose of art is washing the dust of daily life off our souls."

This sentiment captures the essence of creativity as a timeless force—one that transcends borders and speaks to the shared human need for expression, connection, and meaning.

Through history, creativity has acted as a bridge between the tangible and the intangible, the individual and the collective. Whether through storytelling, architecture, music, or art, it continues to define and celebrate the human experience, proving that while cultures may differ, the creative spirit remains a shared and enduring legacy.

A Timeless Journey

Creativity has always been humanity's guiding star, illuminating the path from survival to self-expression, from imagination to cultural identity. It is a force so intrinsic to our nature that it transcends the boundaries of time, geography, and technology. From the crude yet meaningful strokes on cave walls to the breathtaking precision of digital art, creativity is the invisible thread that weaves together the narrative of human history. It is a story that began in the primal needs of our ancestors and continues to evolve with each new generation.

In the beginning, creativity was born out of necessity—a tool for survival. But as humans gained mastery over their environment, it became something more profound: a way to communicate, to connect, and to leave a legacy. Those first cave paintings were not just depictions of daily life; they were early attempts to express something beyond the immediate—a connection to others, a reflection of the world, and perhaps even a question to the unknown.

As humanity progressed, creativity evolved into a reflection of imagination and identity. It became the medium through which cultures told their stories, celebrated their achievements, and questioned their realities. From the architectural marvels of ancient civilizations to the poetic verses of great thinkers, every era has contributed a unique chapter to this timeless journey.

The Renaissance, for example, was a period when creativity redefined what it meant to be human. It bridged the gap between science and art, reason and passion, reality and possibility. Thinkers like Leonardo da Vinci didn't see boundaries; they saw infinite opportunities to explore, create, and innovate. This spirit of exploration is what makes creativity not just a skill, but a defining feature of humanity.

In today's world, creativity has taken on new dimensions, fueled by technology. Digital tools have transformed the way we create, enabling us to reach new heights of precision, efficiency, and expression. Yet, even as the tools evolve, the essence of creativity remains unchanged. It is still about connecting with the human spirit, about asking "what if" and daring to imagine answers.

"Creativity is seeing what everyone else has seen and thinking what no one else has thought,"

said Albert Einstein, capturing the timeless nature of this force.

Creativity isn't just something we do—it is who we are. It is embedded in the way we think, the way we solve problems, and the way we express ourselves. It is the universal language that unites us, even as it allows us to celebrate our individuality. Whether through a song, a story, a painting, or a scientific breakthrough, creativity continues to remind us of our boundless potential.

As we journey through time, we see that creativity is not static; it is a dynamic and ever-evolving force. Each generation inherits the creative spirit of its predecessors, adapting it to the challenges and opportunities of its age. It is a timeless force, connecting us to our past, defining our present, and inspiring our future. Creativity is not just a part of the human experience—it *is* the human experience, a universal story that continues to unfold with every passing moment.

1.2 The Role of Tools in Human Evolution

Creativity and tools have always shared an inseparable bond. Tools are more than mere instruments—they are extensions of human capability, amplifying our ability to shape the world around us. From the crude stone axes of early humans to the sophisticated technologies of the digital age, tools have played a pivotal role in the evolution of creativity, enabling us to express, innovate, and transform. As tools evolved, so did the horizons of human potential, creating milestones that forever changed the way we interact with our environment and each other.

Tools as Extensions of Human Capability

At their essence, tools are embodiments of human ingenuity, transforming thoughts into tangible outcomes. They represent the intersection of necessity and creativity, serving as a bridge between what humans imagine and what they can achieve. The earliest tools, carved from stone, bone, and wood, addressed basic survival needs—making it easier to hunt, gather, build, and protect. Yet, despite their apparent simplicity, these tools were revolutionary. They marked the moment when humans began to see the potential of shaping the world to their advantage, turning challenges into opportunities.

The creation of tools required more than just manual skill; it demanded imagination and foresight. A sharpened stone wasn't merely a cutting implement—it was a leap in understanding cause and effect, a glimpse of how raw materials could be transformed to serve a purpose. Each tool was an acknowledgment of the human capacity to think ahead, to predict outcomes, and to solve problems. For instance, the crafting of a pot to store water wasn't just about utility; it reflected a growing awareness of the future—a recognition that preparation could ease the uncertainties of life.

As tools became more sophisticated, their impact extended beyond physical tasks. They began to influence how humans perceived and interacted with their environment. Tools didn't just change what humans could do; they changed how they thought. The ability to create something functional spurred a deeper engagement with problem-solving and innovation. A simple tool became an agent of transformation, reshaping not only the physical world but also the cognitive landscape of its maker.

This evolution underscores a critical truth: tools are more than extensions of the human hand—they are extensions of the human mind. They amplify our capacity to envision, innovate, and adapt. Over time, tools enabled humans to move from reactive survival to proactive creation, laying the groundwork for civilization. The development of a wheel was not just a technological breakthrough but a conceptual one, opening up new possibilities for movement, trade, and interaction.

The significance of tools lies not just in their utility but in what they represent about the human spirit. They symbolize our innate drive to overcome limitations and expand possibilities. Each tool, from the humblest hammer to the most advanced algorithm, carries with it the essence of human ambition—the desire to shape, refine, and improve. Tools have always been more than objects; they are catalysts of progress, enabling humanity to reach beyond its immediate grasp and unlock the boundless potential of creativity.

Milestones in Creative Tools: From Stone to Silicon

The evolution of tools is a chronicle of humanity's creative milestones, each breakthrough expanding the horizons of what was possible. These tools didn't merely solve problems—they redefined the way humans interacted with their world, enabling leaps in knowledge, innovation, and expression. From the discovery of fire to the rise of artificial intelligence, each stage in this journey has pushed the boundaries of human potential.

The Dawn of Tools: Fire and the Wheel

The discovery of fire marked humanity's first major leap in tool-making. Fire provided warmth, light, and a means to cook food, fundamentally altering the way humans lived and survived. Yet, fire was more than a practical resource; it was a symbol of control over nature, an early indication of humanity's ability to harness elements to serve its needs. The tools used to ignite and contain fire represented the first steps in transforming raw materials into purposeful inventions.

The invention of the wheel, another monumental milestone, revolutionized transportation and trade. No longer bound by the limits of manual labor or geography, humans could now move goods, ideas, and cultures over vast distances. The wheel became a cornerstone for the development of civilizations, enabling growth and interconnectedness on an unprecedented scale. It was a simple yet profound leap that demonstrated how tools could reshape the fabric of society.

The Printing Press: Democratizing Knowledge

Fast-forward to the 15th century, and the printing press emerged as a transformative force. Johannes Gutenberg's invention was not just a technological innovation; it was a cultural revolution. By enabling the mass production of books, the printing press democratized knowledge, breaking the monopoly of information held by the elite. It empowered thinkers, inspired movements, and laid the foundation for the Renaissance—a golden age of art, science, and philosophy.

The printing press proved that tools could be more than functional—they could be catalysts for intellectual and societal transformation. By making information accessible, it fueled a new era of creativity, where ideas could be shared, challenged, and expanded upon at an unprecedented scale.

The Industrial Revolution: Mechanization and Precision

The industrial revolution brought machinery that transformed the way humans created and built. Tools like the steam engine and automated looms revolutionized manufacturing, enabling mass production and unprecedented precision. These advancements weren't just about efficiency—they redefined creativity itself. For the first time, complex ideas could be translated into large-scale creations, from intricate textiles to towering architectural marvels.

Mechanization also gave rise to new disciplines of design and engineering, where creativity met technical expertise. The industrial revolution demonstrated how tools could amplify human ingenuity, turning dreams of scale and complexity into reality.

The Digital Age: From Computers to Artificial Intelligence

The 20th century marked the beginning of the digital age, where silicon chips became the building blocks of a new creative frontier. The computer, initially a tool for calculation, quickly evolved into a platform for design, communication, and innovation. Digital tools transformed every creative discipline, from music and film to architecture and science, allowing humans to push the boundaries of imagination.

The advent of artificial intelligence has further accelerated this trajectory. AI tools now assist in creating art, composing music, and solving complex problems, learning and adapting alongside human creators. They amplify human capabilities, making it possible to achieve in moments what once took years. Yet, these tools also challenge our understanding of creativity, raising questions about the role of human input and the nature of originality.

A Journey Without Limits

From stone tools to silicon chips, each milestone in tool-making reflects humanity's relentless drive to innovate and create. These tools have not only shaped how we interact with the world but also how we think about our place within it. They are reminders that creativity is not bound by the limitations of the present—it is a force that constantly seeks new horizons, propelled by the tools we invent to explore them.

How Tools Have Transformed Human Expression

Tools have always been more than functional instruments; they are the vessels through which human creativity finds its voice. As they evolved, tools fundamentally transformed how we express ourselves, turning raw imagination into tangible, impactful creations. From the first carvings on stone to the immersive experiences of virtual reality, tools have expanded the boundaries of human expression, allowing us to communicate, connect, and innovate in unprecedented ways.

The Beginnings of Expression: Carvings and Symbols

The story of tools transforming expression begins with the earliest humans. Armed with rudimentary tools, they carved symbols into stone, giving rise to the first forms of art and communication. These engravings were more than decorative—they were a leap into storytelling, an extraordinary development that allowed humans to share ideas, document their lives, and preserve their histories. Through these simple yet profound acts, tools enabled early humans to transcend their immediate existence, creating a bridge to future generations.

Expanding Horizons: Music, Architecture, and Painting

As tools advanced, they opened doors to more nuanced forms of expression. The creation of musical instruments introduced a new language of emotion, where melodies and rhythms could convey feelings beyond words. A simple drumbeat, a string plucked, or a flute's gentle note became tools for storytelling, celebration, and mourning, uniting communities through shared experiences.

Architecture became a canvas for grandeur and faith. Tools like chisels, hammers, and pulleys transformed stone and wood into awe-inspiring temples, cathedrals, and monuments. These structures weren't just functional—they were expressions of human aspiration and spiritual devotion, reflecting the dreams and beliefs of entire civilizations.

Painting, too, evolved into a medium for exploring the depths of the human psyche. Brushes and pigments allowed artists to capture not only the world around them but also the emotions and ideas within. From the intricate details of Renaissance masterpieces to the bold abstractions of modern art, tools enabled painters to push the limits of creativity, offering new perspectives on reality.

Revolutionizing Creativity: Photography and Film

The invention of the camera in the 19th century marked a turning point in creative expression. For the first time, humans could capture fleeting moments with precision, preserving them for eternity. Photography transformed how we document life, tell stories, and see the world, providing a new lens through which to view reality.

Film took this transformation further, combining visuals with motion and sound to create immersive narratives. Tools like editing software, sound equipment, and special effects expanded the scope of storytelling, enabling filmmakers to transport audiences to imagined worlds and evoke emotions on an unprecedented scale.

The Digital Renaissance: Unlimited Possibilities

In the digital era, tools like graphic design software, 3D printers, and virtual reality have redefined the possibilities of creative expression. Artists can now sculpt without clay, paint without brushes, and compose music without instruments, using digital platforms that expand the horizons of imagination. These tools allow creators to explore dimensions of art, design, and innovation that were once unimaginable.

Virtual reality, for instance, immerses viewers in entirely new worlds, transforming passive audiences into active participants. 3D printing enables designers to turn digital models into physical objects with unparalleled precision, revolutionizing industries from fashion to engineering. Graphic design software empowers creators to combine text, images, and effects into stunning visuals that captivate and inspire.

The Transformative Power of Tools

As Archimedes once said,

"Give me a place to stand, and I will move the world."

Tools have not only moved the world but have also reshaped how we see it, express it, and interact with it. They have amplified human creativity, enabling us to transcend physical and intellectual limitations and reach new heights of innovation.

Looking back at the milestones in the evolution of tools, it becomes clear that they are not mere utilities—they are catalysts of transformation. From stone carvings to virtual realities, tools have expanded our capacity to create, connect, and innovate. They remind us that while creativity is innate, its possibilities are infinite when paired with the right instruments. Tools are, and always will be, the keys that unlock the full potential of human expression.

1.3 Creativity in the Context of Change

Change is the only constant, and creativity is humanity's most powerful response to it. Whether adapting to shifting environments, embracing technological revolutions, or navigating societal transformations, creativity has proven to be our enduring guide. It is the bridge between the challenges of today and the solutions of tomorrow, thriving in the face of uncertainty and fueling progress across ages.

Adaptability as a Core Element of Creativity

Creativity and adaptability are inseparable, with adaptability serving as the foundation upon which creativity thrives. At its core, adaptability is the ability to navigate and embrace change with ingenuity and resilience. It is the capacity to see challenges as opportunities and to craft innovative solutions that go beyond mere survival, driving progress and evolution.

From the dawn of humanity, adaptability has been the cornerstone of creative thought. Early humans faced harsh and unpredictable environments, relying on their ability to innovate to secure food, shelter, and safety. They crafted tools from stone, bone, and wood, not merely to respond to their immediate needs but to anticipate future challenges. This ability to foresee potential problems and devise solutions transformed their way of life, laying the groundwork for the evolution of civilization.

Adaptability, however, is not just a response to change—it is a proactive embrace of it. Creators often thrive where others falter, seeing beyond the uncertainty to uncover hidden possibilities. For instance, the transition from hunter-gatherer societies to agricultural civilizations marked one of humanity's most significant adaptive leaps. It required people to rethink their relationship with the land, transforming wild landscapes into organized fields capable of sustaining large populations. This shift was not a reaction to immediate necessity but a forward-thinking approach that anticipated the benefits of stable food supplies and long-term settlement.

This dynamic ability to adapt extends far beyond societal changes—it is a hallmark of individual creativity as well. History is filled with examples of artists, scientists, and innovators who flourished because they embraced change rather than resisted it. Leonardo da Vinci, for instance, adapted his creative pursuits to encompass multiple disciplines, from art to engineering, allowing his genius to resonate across centuries. Similarly, scientists like Marie Curie pushed the boundaries of their time, adapting their methods and theories to uncover groundbreaking discoveries.

Adaptability in creativity is also about flexibility—being open to evolving ideas and approaches. Rigidity can stifle innovation, while a willingness to explore, experiment, and adjust leads to breakthroughs. For example, during the industrial revolution, adaptability enabled creative minds to harness emerging technologies, transforming manual craftsmanship into mechanized production. This flexibility didn't just redefine industries; it opened new avenues for art, design, and expression.

In the modern era, adaptability is more critical than ever. The rapid pace of technological advancement demands a creative mindset that can keep up with change and leverage it effectively. Whether it's an artist embracing digital tools, a scientist pivoting research in response to new discoveries, or a business innovator navigating shifting market dynamics, adaptability fuels creativity in every field.

Adaptability ensures that creativity remains relevant and impactful. It allows individuals and societies to evolve alongside new challenges, transforming obstacles into stepping stones for growth. More importantly, it keeps creativity alive as a living, breathing force—one that flows, adapts, and thrives in the face of change. Without adaptability, creativity would stagnate, losing its ability to inspire progress. With it, creativity becomes a powerful catalyst for transformation, proving that resilience and openness are essential elements of human ingenuity.

Creativity in Times of Technological Shifts

Throughout history, technological shifts have been both disruptors and enablers of creativity, forcing humanity to rethink established norms while opening doors to unprecedented opportunities. These transformative periods serve as crucibles for creative innovation, challenging individuals and societies to adapt, experiment, and explore new frontiers of expression and problem-solving.

The Printing Press and the Democratization of Knowledge

One of the earliest examples of a technological shift sparking creativity was the invention of the printing press in the 15th century. Before its arrival, knowledge was confined to hand-copied manuscripts, accessible only to the elite. The printing press disrupted this exclusivity, democratizing knowledge and inspiring an explosion of intellectual and artistic creativity during the Renaissance. It allowed ideas to flow freely, fostering collaboration among thinkers, writers, and artists across Europe. The ability to mass-produce books not only revolutionized education but also fueled the creation of new literary and artistic movements, redefining cultural landscapes.

The Industrial Revolution: Mechanization Meets Innovation

The industrial revolution marked another profound technological shift, mechanizing processes that had traditionally been manual. While some feared that machines would stifle human creativity, they instead inspired new forms of artistic and engineering expression. Factories became sites of innovation, where designers and engineers reimagined production methods, creating intricate textiles, architectural marvels, and revolutionary inventions like the steam engine.

This era also gave birth to industrial design as a discipline, blending functionality with aesthetics. It demonstrated that creativity could thrive even in highly mechanized environments, proving that technology could enhance rather than diminish the human touch.

The Digital Age: Redefining Creativity

In the digital age, technology has propelled creativity into entirely new realms. Tools like artificial intelligence, blockchain, virtual reality, and 3D printing are reshaping how we create, share, and experience art, ideas, and solutions. These technologies challenge traditional notions of creativity, raising questions about authorship, originality, and the role of human input in machine-assisted creation.

Artificial intelligence, for example, has introduced a collaborative dimension to creativity. AI tools can generate visual art, compose music, and even write stories, offering creators a new kind of partnership. Rather than replacing human creativity, these tools amplify it, providing novel ways to experiment and innovate. Similarly, blockchain technology has revolutionized the art world, enabling digital artists to establish provenance and ownership through NFTs (non-fungible tokens), creating entirely new markets for creative work.

Virtual reality and augmented reality have transformed storytelling and design, immersing audiences in experiences that were once the stuff of dreams. Filmmakers can now construct entire worlds using CGI, while architects use virtual models to bring their visions to life before a single brick is laid. These tools expand the boundaries of imagination, allowing creators to explore dimensions of art and innovation that were previously unimaginable.

The Evolution of Digital Art Forms

One of the most striking examples of creativity in the digital age is the evolution of digital art. Where painters once relied on canvas and brushes, today's artists wield tablets and styluses to craft masterpieces with precision and versatility. Musicians compose symphonies using sophisticated software, blending traditional instruments with electronic sounds. Writers refine their ideas with the help of AI-powered tools, transforming the writing process into a seamless collaboration between human and machine.

Even fields traditionally considered outside the realm of creativity, such as engineering and data science, have embraced this transformation. Data visualization, for instance, has become an art form, turning raw numbers into compelling stories. Innovations in software and algorithms have made it possible to model and simulate complex systems, unlocking new ways to design, solve problems, and express ideas.

Creativity Amid Disruption

Technological shifts may disrupt established practices, but they also expand the horizons of creativity. Each new tool or innovation challenges creators to adapt, integrate, and rethink their craft. As humans incorporate these tools into their creative processes, they redefine what it means to create, proving that creativity is not constrained by the medium but enriched by it.

The digital age underscores a timeless truth: creativity is inherently adaptive. It thrives in disruption, drawing strength from the challenges and opportunities posed by technological advancements. While the tools we use may change, the drive to create remains constant, ensuring that creativity continues to evolve alongside humanity's ever-changing landscape.

Why Creativity Remains Timeless

Creativity, at its core, is timeless because it transcends the tools, technologies, and societal norms of any given era. While the mediums through which we create may evolve, the essence of creativity remains unchanged: it is the drive to explore, to question, and to imagine a reality beyond what is immediately visible. Creativity is not tied to any specific method or moment; it is an enduring reflection of the human spirit's relentless quest for meaning, connection, and progress.

What anchors creativity across ages is its universality. The act of creation, whether it involves shaping the first wheel from stone, composing a symphony that stirs the soul, or coding an algorithm that redefines industries, draws on the same fundamental human qualities—curiosity, imagination, and problem-solving. It connects people across generations and cultures, creating a shared lineage of innovation that links the past, present, and future.

Creativity's timeless nature is also rooted in its ability to adapt and endure. It evolves alongside humanity, integrating new tools and techniques without losing its essence. For example, the invention of the printing press didn't replace storytelling—it amplified it, making narratives more accessible and powerful. Similarly, the rise of digital tools hasn't diminished traditional art forms but has expanded their possibilities, giving creators new ways to express and innovate.

This adaptability ensures that creativity not only survives but thrives amid change.

As John F. Kennedy once remarked,

"The problems of the world cannot possibly be solved by skeptics or cynics whose horizons are limited by the obvious realities. We need men who can dream of things that never were."

Creativity embodies this dream—it is the capacity to envision what could be and to bring those visions to life, even when they defy conventional boundaries.

Creativity's endurance also stems from its deeply personal and emotional roots. It is driven by an innate need to understand, communicate, and contribute. For some, creativity is a means of exploring and expressing their identity; for others, it is a way to solve pressing challenges or improve the lives of others. This versatility makes it relevant to every individual and era, ensuring its continued resonance regardless of the external environment.

Moreover, creativity is not just about responding to change—it is about shaping it. It drives innovation, challenges the status quo, and lays the foundation for progress. From the grand designs of Leonardo da Vinci to the revolutionary theories of Einstein, creativity has consistently been the force that propels humanity forward, turning dreams into reality and problems into possibilities.

As humanity faces an increasingly complex and interconnected world, creativity remains our constant. It bridges disciplines, cultures, and generations, reminding us of our shared potential to imagine and innovate. It is not limited by tools, defined by technology, or confined by tradition.

Instead, it thrives in the ever-changing landscape of human experience, proving that the act of creation is as eternal as the human spirit itself.

Ultimately, creativity is not just a response to the world—it is how we shape it. It ensures that, no matter what the future holds, the human spirit will continue to dream, create, and inspire. This is why creativity remains timeless: because it is, and always will be, the essence of what it means to be human.

Summary

This chapter traces the journey of creativity through human history, showcasing how it has evolved from survival-based problem-solving to a driving force behind art, culture, and technological progress. Starting with the invention of tools and fire, it highlights pivotal moments such as the agricultural revolution, the Renaissance, and the industrial age, illustrating how human ingenuity has continuously adapted to new challenges and opportunities.

The chapter emphasizes that creativity has always been a response to change—whether in the form of cultural shifts, technological breakthroughs, or societal needs. It also sets the stage for the modern era, where AI and machine learning represent the latest frontier in human innovation. The narrative underscores the timeless nature of creativity while inviting readers to reflect on their role in shaping its future.

Part I: Creativity and Tools – A Historical Perspective

"Throughout history, tools have been the threads that weave human ingenuity into reality. They are not the creators but the catalysts, enabling us to transform ideas into timeless legacies."
— Sandeep Chavan

2: From Brush to Algorithm

The story of human creativity is inseparable from the tools we have used to express, build, and innovate. From the first sharpened stones to the intricate algorithms of modern artificial intelligence, tools have been the silent partners in our creative journey. They have not only amplified our abilities but also redefined the way we perceive and interact with the world. Each tool, whether as simple as a brush or as complex as a machine-learning algorithm, carries with it the potential to expand the boundaries of human imagination. Yet, the rise of new tools has often been met with skepticism and resistance, as history reveals a recurring tension between the fear of change and the embrace of progress. This chapter explores how tools have shaped human innovation and examines the lessons we can draw from past apprehensions toward technological advancements.

"Creativity is not confined by the tools we use—it is the eternal flame that fuels our quest to dream, build, and transform."

2.1 How Tools Shaped Human Innovation

The journey of human innovation is intricately tied to the tools we've created. Each tool, born out of necessity or imagination, has propelled humanity forward, enabling us to solve problems, improve our lives, and expand our creative horizons. From the first primitive implements to the advanced algorithms of today, tools have been at the heart of our progress, transforming not only what we create but also how we think.

The First Tools: Necessity Breeding Ingenuity

The story of human innovation begins with the simplest tools—stones, bones, and sticks—crafted from the materials readily available in nature.

These early implements, while primitive by today's standards, marked the first significant leap in humanity's ability to shape its environment. They were extensions of human hands, enabling our ancestors to perform tasks that were otherwise impossible. From hunting and gathering to building rudimentary shelters and defending against predators, these tools played a critical role in ensuring survival.

What set these early tools apart was their transformative nature. They were not mere objects; they were solutions born out of necessity. A pointed stick or a sharpened stone wasn't just a means to an end—it was a testament to the ingenuity and resourcefulness of early humans. By observing their environment and understanding the properties of materials around them, they were able to imagine and create tools that fundamentally altered their interaction with the world.

The first tools also marked a cognitive shift. Crafting them required foresight, planning, and an understanding of cause and effect. For example, sharpening a stone to create a cutting edge was more than an act of physical labor—it demonstrated an ability to anticipate future needs and prepare accordingly. This leap in thinking reflected a deeper connection between human creativity and survival, where innovation became an adaptive strategy in the face of challenges.

These tools were simple yet revolutionary. A sharpened stone, used for cutting meat or scraping hides, was a breakthrough in efficiency and utility. A sturdy branch, honed into a spear, extended the reach and power of the human arm, transforming hunting techniques and opening up new opportunities for sustenance. Each innovation, no matter how basic, represented a step toward mastering the environment.

What makes these early tools particularly remarkable is the imagination involved in their creation. Where others might have seen a rock or a stick, early humans saw potential—a means to solve problems and improve their circumstances. This ability to see beyond the immediate and envision new possibilities is what set humanity apart and laid the groundwork for future advancements.

The impact of these tools extended far beyond their immediate utility. They were the building blocks of a creative legacy, setting a precedent for problem-solving and innovation that continues to define humanity. Each sharpened stone, each crafted implement, carried within it the seeds of invention, proving that necessity is indeed the mother of ingenuity.

In creating these tools, early humans not only adapted to their environment but also began to shape it. They demonstrated an understanding of materials, a willingness to experiment, and an ability to learn from failure—all hallmarks of creativity that would drive progress for millennia. These first tools may have been born out of necessity, but they sparked a journey of innovation that would eventually lead to the marvels of modern technology.

Catalysts for Progress: Tools That Transformed Civilizations

As human societies transitioned from small, nomadic groups to complex civilizations, the role of tools became increasingly significant. These innovations were no longer designed solely for individual survival but to meet the growing demands of entire communities. Tools like the wheel, the plow, and irrigation systems were not only practical—they were transformative, laying the foundation for the rise and flourishing of civilizations.

Revolutionizing Agriculture

One of the most profound ways tools shaped civilization was through agriculture. The plow, for instance, revolutionized farming by making it possible to cultivate land more efficiently and on a larger scale. Paired with irrigation systems that redirected water to arid regions, these tools ensured stable food supplies, allowing populations to grow and settle in one place. Agriculture, sustained by these innovations, became the backbone of early civilizations, enabling the rise of cities, governance, and trade networks.

These agricultural tools were not just mechanical devices; they represented a new way of thinking about the environment. The plow transformed the land into a resource that could be harnessed for long-term prosperity, while irrigation systems turned the unpredictable flow of rivers into dependable lifelines. Together, these tools demonstrated humanity's ability to adapt to and shape its surroundings, fostering stability and progress.

The Wheel and the Expansion of Movement

The invention of the wheel was another milestone that reshaped human society. Initially developed for pottery, it quickly evolved into a fundamental tool for transportation. The wheel enabled the creation of carts and wagons, drastically improving the efficiency of moving goods and people. Trade routes expanded, connecting distant regions and facilitating cultural exchange.

This newfound mobility wasn't just about economics—it was about connection. The wheel allowed for the sharing of ideas, goods, and innovations, knitting together the fabric of early civilizations. It became a symbol of progress, representing the ability to transcend geographic limitations and foster interdependence among communities.

Engineering Marvels: Levers, Pulleys, and Architecture

In the realm of construction, tools like levers and pulleys became indispensable. They made it possible to build monumental structures that still stand as testaments to human ingenuity. The pyramids of Egypt, for example, required sophisticated knowledge of mechanics and the ability to organize labor on an unprecedented scale. Similarly, the aqueducts of Rome, powered by a deep understanding of physics and engineering, brought water to urban centers, transforming the way people lived.

These tools weren't just practical implements; they were enablers of ambition. They allowed civilizations to dream bigger, to create structures that embodied cultural, religious, and political ideals. The ability to move massive stones, erect towering columns, and redirect water over long distances was a reflection of humanity's growing mastery over the natural world.

Catalysts for Growth and Exchange

What made these tools truly transformative was their ripple effect. By improving agriculture, transportation, and construction, they drove economic growth, technological advancement, and cultural exchange. Stable food supplies freed people to pursue other activities, leading to the development of crafts, trade, and governance. Improved transportation networks facilitated the flow of goods and ideas, sparking innovation and fostering relationships between distant regions.

These tools also played a critical role in advancing technology. The wheel and axle, for example, became the foundation for countless mechanical devices, while agricultural innovations paved the way for advancements in metallurgy and machinery. Each tool built upon the successes of the last, creating a cumulative effect that propelled societies forward.

Instruments of Civilization-Building

Above all, these tools symbolized humanity's ability to think beyond immediate needs and envision solutions that benefited entire communities. They transformed raw potential—whether it was untamed land, unconnected regions, or unorganized resources—into tangible outcomes that shaped the course of history. The wheel, the plow, levers, and irrigation systems weren't just practical devices; they were instruments of civilization-building, embodying the ingenuity, ambition, and collaborative spirit that define humanity.

These catalysts for progress remind us that tools are far more than functional objects. They are expressions of creativity and foresight, capable of transforming societies and laying the groundwork for future advancements. Each innovation, no matter how simple it may seem in hindsight, carries with it the power to reshape the world, demonstrating that tools are as much about vision as they are about utility.

From Manual to Mechanized: The Evolution of Tool-Making

The industrial revolution marked a transformative era in the history of tools, fundamentally altering how humans approached creation and production. Manual craftsmanship, which had long been the cornerstone of human ingenuity, gave way to mechanized processes powered by groundbreaking innovations like the steam engine and automated machinery. These tools not only made work faster and more efficient but also redefined the boundaries of what was possible.

The Leap to Mechanized Production

Mechanization revolutionized industries by replacing labor-intensive manual processes with machine-driven systems. Factories equipped with steam-powered machinery became the epicenters of production, churning out goods at scales and speeds that were previously unimaginable. This shift allowed for the mass production of textiles, iron, and consumer goods, making products more affordable and accessible to wider populations.

Beyond efficiency, mechanization enabled precision that was difficult to achieve with manual tools. Machines could produce intricate components repeatedly with exact measurements, laying the groundwork for industries like automotive and aerospace manufacturing. The advent of interchangeable parts further streamlined production, standardizing processes and enabling rapid assembly in industries ranging from weaponry to home goods.

Transforming Creativity at Scale

Mechanization wasn't just about making existing processes faster—it opened up entirely new possibilities for human creativity. The ability to mass-produce goods transformed not only commerce but also culture. Tools like the printing press, which had already revolutionized knowledge dissemination, became even more efficient, making books, newspapers, and pamphlets more widely available. This democratization of information fueled education, scientific inquiry, and social movements, reshaping societies.

Advances in engineering, enabled by mechanized tools, allowed humans to tackle ambitious projects on a grand scale. Railways connected distant cities, fostering trade and communication. Bridges spanned vast rivers, linking communities and enabling economic growth. Skyscrapers redefined urban landscapes, becoming symbols of progress and ingenuity. Each of these feats was made possible by tools that combined power, precision, and innovation.

The Rise of Industrial Design

Mechanization also gave birth to industrial design, blending functionality with aesthetics. No longer confined to mere utility, tools and products became symbols of innovation and craftsmanship. Designers began to focus not only on how tools performed but also on how they looked and felt, creating a harmony between form and function.

The integration of design into mechanized production was evident in everything from furniture to vehicles. For instance, the streamlined shapes of early automobiles and the ornate details of cast-iron railings reflected a new era where tools and products were crafted to be both practical and visually appealing. This emphasis on design elevated the status of manufactured goods, turning them into cultural icons and status symbols.

Reshaping Society and Urban Landscapes

Mechanized tools also reshaped human settlements, driving urbanization and industrial expansion. Factories became the heart of growing cities, attracting workers and spurring the development of infrastructure. Urban landscapes transformed with the construction of railways, bridges, and towering buildings, all made possible by advances in mechanized engineering.

The impact extended beyond cities. Mechanization revolutionized agriculture through innovations like the mechanical plow and threshing machines, increasing food production and enabling rural populations to thrive. These tools reduced the physical burden on farmers and boosted productivity, paving the way for modern agribusiness.

Paving the Way for Modern Advancements

The mechanization of tools during the industrial revolution set the stage for future technological advancements. It marked humanity's shift from small-scale, handcrafted ingenuity to large-scale, systematized creativity. The principles of mechanization—efficiency, precision, and scalability—became the foundation for modern industries, influencing everything from assembly lines to digital manufacturing.

This era demonstrated that tools are more than implements; they are enablers of transformation. Mechanization allowed humans to transcend limitations, turning ambitious dreams into tangible realities. It redefined the relationship between creator and creation, proving that the evolution of tools is not just a story of progress but a testament to the boundless potential of human ingenuity.

The Digital Leap: Algorithms as Creative Tools

The advent of the digital age represents a revolutionary leap in the evolution of tools. Unlike their physical predecessors, which primarily extended human physical capabilities, digital tools operate in the abstract realm of data and code. Despite their intangible nature, their impact on how we create, solve problems, and interact with the world is deeply transformative, reshaping industries, redefining creativity, and expanding the horizons of human potential.

Algorithms as Engines of Innovation

At the heart of the digital revolution are algorithms—sequences of instructions that process vast amounts of data to produce desired outcomes. These computational tools have become the engines of innovation, powering applications across diverse fields. From creating intricate digital art to composing symphonies, from designing predictive models in engineering to unlocking the secrets of the human genome, algorithms are opening new dimensions of creativity and discovery.

Unlike traditional tools, algorithms don't just execute tasks—they learn, adapt, and optimize. For instance, generative algorithms in artificial intelligence can analyze patterns and generate creative outputs, such as visual art or music, that rival human ingenuity. Machine learning algorithms enhance this capability by refining their processes over time, allowing them to produce increasingly sophisticated results. This adaptability transforms algorithms into collaborators in the creative process rather than mere instruments.

Expanding the Boundaries of Creation

The digital leap has enabled creators to work with tools that vastly extend the scope and scale of their imagination. Graphic design software allows artists to manipulate images with precision and versatility, enabling them to experiment with ideas that would be impossible in traditional mediums. Virtual reality platforms immerse users in fully realized environments, empowering architects, designers, and storytellers to craft experiences that engage multiple senses.

Artificial intelligence has further expanded these possibilities. AI-assisted tools can analyze trends, generate ideas, and even offer creative suggestions, acting as partners in the creative process. For instance, a writer might use AI to brainstorm plotlines or refine prose, while a musician might collaborate with an AI system to compose a unique piece of music. These tools free creators from the constraints of repetitive or time-consuming tasks, allowing them to focus on higher-level thinking and innovation.

Redefining Creativity with Digital Tools

The digital leap has not only enhanced the tools at our disposal but also redefined the very nature of creativity. In traditional contexts, creativity was often seen as an individual or collective effort, deeply rooted in human intuition and expression. However, with the rise of digital tools, creativity has become a collaborative process between humans and machines. Algorithms are now active participants, contributing ideas, analyzing data, and optimizing workflows in ways that were once unimaginable.

For example, in the field of design, algorithms can generate countless variations of a concept, enabling designers to explore a broader range of possibilities. In engineering, they can simulate and optimize complex systems, reducing development times and costs. In art and entertainment, tools like neural networks and deep learning models are creating entirely new genres of creative expression, from AI-generated paintings to virtual actors in movies.

Collaboration, Not Replacement

A common misconception about digital tools, particularly artificial intelligence, is that they aim to replace human creativity. In reality, these tools are designed to augment human capabilities, enhancing what we can achieve rather than diminishing our role. By automating repetitive or labor-intensive tasks, algorithms allow creators to focus on the aspects of their work that require intuition, emotion, and innovation.

For instance, filmmakers use AI to edit footage, generate visual effects, and even predict audience preferences. Scientists employ algorithms to analyze massive datasets, uncovering patterns and insights that would be impossible to identify manually. In these contexts, the human mind remains at the helm, guiding the creative process and making the critical decisions that shape the final output.

Transforming the Definition of Tools

The digital age has fundamentally transformed what we understand tools to be. They are no longer mere instruments that extend our physical reach; they are dynamic systems that amplify our intellectual and creative capacities. Algorithms and digital platforms are not static—they evolve, learn, and adapt alongside their users, creating a symbiotic relationship that continuously pushes the boundaries of what can be achieved.

This transformation has also broadened access to creativity. Digital tools democratize creation by making advanced capabilities available to individuals and small teams, empowering more people to innovate and express themselves. Whether it's an independent filmmaker using affordable editing software or a student exploring virtual reality for the first time, the digital leap has ensured that creativity is no longer confined to those with specialized training or resources.

The digital leap represents more than just an evolution in tools—it is a revolution in how we create, collaborate, and think. Algorithms and computational technologies have become integral to the creative process, transforming abstract ideas into tangible realities with speed and precision. As active participants in creation, these tools are reshaping how we define creativity itself, proving that the human imagination, when paired with the power of algorithms, is boundless.

Tools: Reflections of Human Ingenuity and Ambition

Tools have always been more than mere implements; they are profound reflections of human ingenuity, ambition, and our relentless quest for progress. Each tool, whether a simple stone axe or a complex algorithm, tells a story of creativity and problem-solving. It represents a moment in time when humanity looked at its challenges and chose to imagine something better—something that could extend our capabilities, enhance our lives, and reshape our understanding of the world.

The simple stone axe, one of humanity's earliest tools, wasn't just a piece of sharpened rock—it was a leap in thought. It symbolized foresight and adaptability, showing an ability to see beyond immediate survival needs and envision tools that could provide long-term solutions. This early innovation marked the beginning of a transformative journey, where each tool built upon the ingenuity of the last, propelling us toward increasingly complex and ambitious creations.

As tools evolved, so did humanity. The wheel revolutionized transportation and trade, allowing societies to connect, share, and grow. The printing press democratized knowledge, breaking down barriers and enabling ideas to flow freely across borders. The steam engine powered the industrial revolution, reshaping economies and altering the very fabric of daily life. Each of these milestones reflects not only the tool itself but the ambition of those who dared to imagine its potential.

In the modern era, tools like algorithms and artificial intelligence represent the latest chapter in this story. These digital tools operate in abstract realms of data and logic, yet their impact is tangible, transforming industries, enhancing human creativity, and solving problems at scales previously unimaginable. They are the embodiment of our ambition to push boundaries, explore the unknown, and innovate at the edge of possibility.

The relationship between tools and humanity is symbiotic. By shaping tools, we have simultaneously shaped ourselves. Tools have challenged us to think differently, adapt to new paradigms, and stretch the limits of our imagination. They have influenced how we work, communicate, and even how we define creativity itself. As we craft tools, they, in turn, redefine what we can achieve, becoming extensions of our intellect and ambition.

This interplay between tools and humanity underscores a fundamental truth: creativity and innovation are inextricably linked to the tools we wield. They are not just products of necessity but expressions of our deepest aspirations. Each tool, from the simplest to the most sophisticated, stands as a testament to our drive to transform challenges into opportunities and possibilities into realities. In shaping our tools, we are, ultimately, shaping the future of our species.

2.2 Lessons from History: Skepticism Towards New Tools

Throughout history, new tools and technologies have often been met with skepticism and resistance. Despite their potential to revolutionize societies and enhance human capabilities, they have frequently been seen as threats to established norms, values, and livelihoods. This recurring pattern reveals much about human nature: our simultaneous fear of and fascination with change. Examining these moments of resistance offers valuable lessons about our relationship with innovation and how we can better navigate technological disruptions in the future.

The Fear of Change: Early Resistance to Innovation

Change, while a driving force for progress, has always been accompanied by resistance. New tools and technologies often challenge established norms, threatening the comfort of the familiar and provoking fears of disruption. This resistance is not simply about the tools themselves but reflects deeper anxieties about the impact of innovation on livelihoods, traditions, and identities.

Bronze Tools: A Revolutionary Disruption

One of the earliest examples of such resistance can be seen in the transition from stone to bronze tools during the Bronze Age. Societies reliant on stone implements faced upheaval as bronze, a stronger and more versatile material, began to replace traditional tools. This shift required new skills, such as metalworking, and access to resources like copper and tin, which were not readily available to all. Those unable to adapt or access these resources feared being left behind, their way of life rendered obsolete.

The introduction of bronze tools also created social and economic disparities, as those who mastered the new technology gained advantages over those who clung to the old. This dynamic fueled tensions and resistance, as people struggled to reconcile the benefits of progress with the challenges of adapting to change.

Agricultural Innovations and Farmer Resistance

The development of the plow marked another significant leap in tool innovation, transforming agriculture and enabling large-scale farming. However, this advancement was not universally embraced. Traditional farmers, accustomed to hand tools and simpler methods, often resisted the plow, fearing that it would disrupt their established practices.

The plow required different techniques, greater physical effort, and sometimes new types of land ownership or labor arrangements. For many, these changes threatened the stability of their livelihoods and communities. Resistance was particularly strong in societies where agricultural practices were closely tied to cultural identity, as the plow represented not just a technical shift but a challenge to deeply ingrained ways of life.

Artisans vs. Mechanized Tools

Artisans and craftsmen, whose skills had been honed over generations, often viewed mechanized tools as existential threats. When looms and spinning machines began to replace traditional textile production methods during the early industrial era, skilled weavers and spinners protested vehemently. These tools, which could produce goods more quickly and cheaply, were seen as devaluing the craftsmanship and artistry that defined their work.

This resistance was not merely economic—it was deeply personal. For artisans, their tools were an extension of their identity, and mechanized processes threatened to strip away the individuality and pride inherent in their craft. The fear of being replaced by machines was as much about losing a sense of purpose as it was about losing income.

Scribes and the Printing Press

The introduction of the printing press in the 15th century provoked similar resistance from scribes, who had long been the custodians of written knowledge. Hand-copying texts was not just a profession but a revered art form, requiring patience, skill, and dedication. The printing press, capable of mass-producing books, threatened to render this painstaking labor obsolete.

Scribes feared not only for their livelihoods but also for the quality and sanctity of written works. They worried that printed texts would lack the beauty and accuracy of hand-copied manuscripts, diluting the value of knowledge itself. Their resistance highlights the emotional and cultural dimensions of the fear of change, as new tools were seen not just as innovations but as threats to cherished traditions.

A Universal Pattern of Resistance

These examples reveal a universal pattern: resistance to innovation is often rooted in a fear of the unknown and a reluctance to abandon the familiar. Practical concerns about obsolescence, economic disparity, and the loss of skills are compounded by deeper anxieties about cultural and personal identity. New tools, by their very nature, disrupt established practices and force societies to confront change, often provoking a defensive response.

However, these fears, while understandable, often underestimate the potential of innovation to enrich and expand human capabilities. The resistance of early societies to bronze tools, farmers to the plow, artisans to mechanization, and scribes to the printing press eventually gave way to acceptance as the benefits of these tools became evident. Over time, these innovations not only improved efficiency and productivity but also created new opportunities, skills, and cultural expressions.

Lessons for the Present

The fear of change is an intrinsic part of the human experience, but history shows that innovation, despite initial resistance, often leads to progress. Understanding this dynamic can help us navigate modern technological disruptions more effectively. By addressing the concerns of those impacted by change, fostering dialogue, and emphasizing the potential for innovation to complement rather than replace human skills, we can ease the transition and ensure that progress benefits all.

The Printing Press Paradox: Knowledge Expansion vs. Distrust

The invention of the printing press by Johannes Gutenberg in the 15th century stands as one of the most revolutionary advancements in human history. It fundamentally altered the way knowledge was created, shared, and preserved, democratizing access to information on an unprecedented scale. By enabling the mass production of books, the printing press shattered the barriers of exclusivity that had long surrounded knowledge, making it accessible to a far broader audience than ever before. Yet, despite its transformative potential, the printing press was met with significant resistance, revealing a paradoxical relationship between innovation and trust.

A Catalyst for Knowledge Expansion

Before the printing press, books were laboriously copied by hand, often taking months or even years to complete. This painstaking process made books rare and expensive, limiting their availability to elites such as clergy, scholars, and royalty. Gutenberg's movable-type printing press changed this dynamic entirely, allowing for the rapid and inexpensive reproduction of texts. This breakthrough democratized knowledge, empowering individuals with access to literature, religious texts, and scientific discoveries.

The printing press sparked an intellectual awakening, fueling movements like the Renaissance and the Reformation. For the first time, ideas could spread widely and quickly, transcending geographic and cultural boundaries. Literacy rates began to rise as books became more affordable, enabling ordinary people to engage with knowledge that had previously been out of reach. The printing press was not just a technological innovation—it was a social and cultural revolution.

Distrust and Opposition from Traditional Gatekeepers

Despite its obvious benefits, the printing press faced significant opposition from established authorities. Religious leaders, for instance, viewed the press with suspicion and alarm. For centuries, the Church had maintained strict control over the production and interpretation of sacred texts. Hand-copied manuscripts were meticulously overseen to ensure theological consistency, preserving the Church's authority over religious knowledge.

The printing press disrupted this monopoly by making it possible to produce and distribute texts outside the control of the Church. Religious leaders feared that mass production would lead to misinterpretations of scripture, heretical ideas, and challenges to their doctrinal authority. These fears were not unfounded; the spread of Martin Luther's 95 Theses, a catalyst for the Protestant Reformation, was enabled by the printing press.

Scribes, too, resisted the new technology. For generations, they had devoted their lives to the meticulous craft of hand-copying manuscripts, a skill that required years of training and immense dedication. The printing press threatened to render this labor-intensive craft obsolete. For scribes, the press was not merely a tool—it represented the erosion of a deeply respected tradition and the loss of livelihoods tied to it.

Fears of Misinformation and Dilution of Knowledge

Beyond economic and theological concerns, intellectuals and gatekeepers of knowledge worried about the potential for the printing press to spread misinformation. In a world accustomed to the slow, curated dissemination of ideas, the rapid production of texts seemed chaotic and unregulated. Critics argued that mass-produced books might prioritize quantity over quality, diluting the sanctity of knowledge and spreading errors or sensationalism.

These anxieties reflect a timeless tension that accompanies major technological advancements: the fear that increased accessibility might compromise the integrity or value of the information being shared. The printing press, by placing knowledge in the hands of the many, challenged long-held assumptions about who should have the right to produce and consume information.

The Paradox of Liberation and Destabilization

The printing press paradox lies in its dual role as both a liberator and a destabilizer. On the one hand, it expanded the horizons of human understanding, enabling the rapid spread of ideas that fueled progress in science, education, and culture. On the other hand, it disrupted established systems of control, raising concerns about the loss of tradition, authority, and reliability.

Despite this resistance, history has shown that the benefits of the printing press far outweighed its perceived risks. It laid the groundwork for the modern world, revolutionizing communication and accelerating the pace of innovation. It proved that while new tools may disrupt the status quo, they often pave the way for profound societal advancements.

Lessons for Modern Innovations

The story of the printing press offers valuable insights for understanding resistance to modern technologies. Much like the press faced skepticism in its time, today's digital tools—such as artificial intelligence and blockchain—are met with similar fears about misinformation, loss of control, and the erosion of traditional practices. However, as history demonstrates, innovations that initially provoke distrust often become indispensable over time, shaping new norms and expanding possibilities.

The printing press paradox serves as a reminder that while innovation can be unsettling, it is also the catalyst for progress. By addressing the concerns of those affected and ensuring that new tools are used responsibly, we can navigate the challenges of technological change and unlock its transformative potential for the betterment of society.

Industrial Revolution Doubts: Machines vs. Human Craftsmanship

The industrial revolution marked a transformative era in human history, ushering in mechanization on an unprecedented scale. From textiles to transportation, industries underwent sweeping changes as machines replaced manual labor, dramatically increasing productivity and efficiency. However, this era of rapid progress was not universally celebrated. It was accompanied by widespread skepticism, fear, and resistance, particularly from skilled artisans and workers whose traditional livelihoods were directly threatened by the advent of mechanized production.

The Threat to Skilled Artisans and Labor

At the heart of the opposition to mechanization was the perceived threat to skilled craftsmanship. For generations, artisans had honed their skills, creating products imbued with individuality and artistry. The introduction of machines, such as power looms and spinning frames, drastically reduced the need for manual craftsmanship, replacing intricate handwork with faster, standardized production. Workers feared that their unique skills, once highly valued, would become obsolete in a world dominated by machines.

This fear materialized in protests, most notably by the Luddites, a group of 19th-century English textile workers. The Luddites became emblematic of resistance to industrial change, destroying machinery in acts of defiance against what they saw as the dehumanization of their craft. Their protests were not merely about the loss of jobs but also about the erosion of pride, identity, and community tied to traditional artisanal work.

Concerns About Dehumanization and Quality

Critics of the industrial revolution also raised concerns about the broader implications of mechanization. They argued that machines risked dehumanizing labor, reducing workers to mere cogs in an impersonal industrial system. Unlike the intimate connection artisans had with their craft, factory work often involved repetitive, monotonous tasks with little room for creativity or personal expression. This shift not only altered the nature of work but also had profound social and psychological impacts on workers.

There was also anxiety about the quality of goods produced by machines. Critics feared that mass production would prioritize quantity over quality, flooding markets with standardized products that lacked the uniqueness and durability of handcrafted items. The individuality and cultural significance of artisanal goods seemed at odds with the uniformity of machine-made products, leading to concerns about the loss of aesthetic and cultural values.

The Economic and Social Impact

While the concerns about mechanization were valid, the industrial revolution also brought undeniable benefits. Mechanized production significantly increased efficiency, lowering the cost of goods and making them accessible to a broader audience. Textiles, once luxury items, became affordable to the working class, improving living standards and creating new markets.

Moreover, the industrial revolution spurred economic growth and urbanization. Factories became hubs of innovation, driving technological advancements and creating jobs in emerging industries. While some traditional roles disappeared, new opportunities arose in areas such as machine operation, engineering, and trade. These changes, though disruptive, laid the groundwork for the modern economy.

Lessons from Resistance

The resistance to mechanization during the industrial revolution underscores the tension between preserving tradition and embracing innovation. It highlights the human cost of technological progress and the need to address the fears and challenges faced by those directly impacted. Workers' anxieties about dehumanization, job displacement, and the loss of craftsmanship were not simply reactions to change—they were calls for a more equitable and thoughtful approach to progress.

These lessons remain relevant today. As we navigate the rise of automation and artificial intelligence, parallels can be drawn to the industrial revolution. Just as mechanization reshaped industries, modern technologies are transforming the nature of work, sparking similar fears about job loss and the erosion of traditional skills. Addressing these concerns requires balancing innovation with empathy, ensuring that progress benefits as many people as possible.

Striking a Balance Between Tradition and Innovation

The industrial revolution ultimately proved that while technological change can disrupt established systems, it also creates new opportunities for growth and advancement. The challenge lies in managing this transition in a way that respects the contributions of traditional practices while embracing the potential of innovation. By providing training, fostering inclusivity, and prioritizing human well-being, societies can navigate technological shifts more effectively.

The story of mechanization during the industrial revolution is a reminder of the complexities of progress. It shows that while skepticism and resistance are natural responses to change, innovation, when guided by thoughtful policies and practices, can lead to a better future. Balancing the preservation of human craftsmanship with the benefits of mechanization remains a timeless challenge, one that continues to shape the relationship between humanity and its tools.

Digital Disruption: AI, Automation, and the Debate on Creativity

In the digital age, the rapid rise of artificial intelligence (AI) and automation has ushered in a transformative era, sparking profound debates about their implications for creativity, employment, and societal structures. These advanced technologies have demonstrated their potential to generate art, compose music, write stories, and solve complex problems, challenging traditional notions of human creativity and ingenuity. As algorithms increasingly influence creative and professional processes, the tension between their promise and their perceived risks has become a defining feature of the modern technological landscape.

AI's Capabilities and the Redefinition of Creativity

AI-powered tools have showcased remarkable abilities in creative domains, from crafting realistic paintings to composing symphonies and generating written content. These systems operate by analyzing vast datasets, identifying patterns, and producing outputs that often rival human efforts. Generative AI models like GPT and DALL-E have demonstrated their capacity to create original artwork, write compelling narratives, and even suggest innovative solutions to complex challenges.

However, the nature of this creativity has raised philosophical questions. Is AI truly creative, or is it merely mimicking patterns derived from human input? Creativity has traditionally been viewed as a deeply human trait, rooted in emotion, intuition, and the ability to connect seemingly unrelated ideas. AI, by contrast, processes information without consciousness or emotional intent, leading to debates about whether its outputs can be considered genuine acts of creation.

Concerns from Skeptics: The Erosion of Human Roles

Critics of AI and automation argue that these technologies pose significant risks to human creativity and employment. In creative industries, there is growing concern that AI could devalue traditional skills and replace human creators. For example, AI-generated art and music have sparked fears among artists and composers that their unique talents may become obsolete in a world where machines can produce similar outputs at lower costs.

Moreover, there are ethical concerns about AI-generated content, particularly regarding issues of authorship and originality. Who owns the rights to a painting created by an algorithm? Can a machine truly be credited with originality, or does its output inherently belong to the developers who created it? These questions highlight the complexities of integrating AI into creative fields, where the value of work often lies in its personal and emotional resonance.

Bias is another critical issue. AI systems are only as unbiased as the data they are trained on, and there have been instances where algorithms have inadvertently reinforced stereotypes or produced culturally insensitive outputs. Such incidents raise concerns about the ethical implications of relying on AI for creative and decision-making processes.

Proponents' Perspective: Enhancing Human Creativity

While skeptics emphasize the risks, proponents of AI and automation highlight their potential to enhance rather than replace human creativity. They argue that AI tools can act as collaborators, freeing creators from repetitive or mundane tasks and enabling them to focus on higher-level thinking and innovation. For instance, writers can use AI to generate ideas or refine drafts, while designers can leverage algorithms to explore multiple variations of a concept quickly.

In fields like architecture and engineering, AI-powered tools enable professionals to simulate complex systems, optimize designs, and solve problems more efficiently. These technologies do not diminish human input but rather amplify it, providing creators with powerful resources to explore new possibilities and push boundaries.

Proponents also point out that AI democratizes creativity by making advanced tools accessible to a broader audience. With AI-assisted platforms, individuals without formal training in art, music, or writing can produce professional-quality work, fostering greater inclusivity and diversity in creative fields.

The Debate Reflects Broader Tensions

The debate surrounding AI and automation reflects a broader tension between embracing innovation and preserving established practices. Similar to the resistance faced by mechanization during the industrial revolution, skepticism toward AI stems from concerns about cultural and economic disruption. However, history shows that technological advancements often lead to new opportunities and redefine roles rather than eliminating them outright.

For instance, while photography was initially seen as a threat to traditional painting, it eventually became a distinct art form in its own right. Similarly, AI-generated content may coexist with human creativity, carving out a complementary role rather than a competing one. The challenge lies in navigating this transition thoughtfully, addressing the concerns of those impacted while maximizing the potential benefits of these technologies.

The rise of AI and automation underscores the need to balance innovation with ethical considerations and human values. By fostering collaboration between humans and machines, addressing biases, and ensuring fair use of AI-generated content, society can embrace these tools as catalysts for progress. The ongoing debate serves as a reminder that while new technologies may disrupt the status quo, they also offer opportunities to expand the horizons of creativity and redefine what it means to be innovative in the modern age.

Lessons from History

History offers a wealth of examples that illuminate humanity's complex relationship with innovation. Time and again, skepticism toward new tools and technologies has emerged as a natural response to change. This skepticism often arises from a fear of the unknown and a reluctance to abandon the familiar. Whether it was the artisans protesting mechanization during the industrial revolution, the scribes resisting the printing press, or the skepticism toward artificial intelligence today, the root of resistance lies in uncertainty—uncertainty about the future, the value of established practices, and the place of humanity in an evolving world.

Skepticism as a Reflection of Human Nature

The resistance to new tools is not merely a rejection of technology but a deeply human reaction to disruption. Tools are more than functional implements—they are intertwined with our identities, cultures, and ways of life. For artisans, the loom wasn't just a machine; it symbolized the loss of a craft handed down through generations. For scribes, the printing press wasn't just a device; it represented a threat to their role as guardians of knowledge. In each case, the fear was not just about change itself but about what that change would mean for their place in the world.

This skepticism is a natural byproduct of the human tendency to value continuity and tradition. Change, especially when driven by unfamiliar tools, can feel like a challenge to stability and a disruption to deeply ingrained systems. However, as history has shown, while skepticism may be a barrier in the short term, it rarely halts the march of progress. Instead, it serves as a checkpoint, prompting society to question, refine, and adapt new technologies in ways that make them more equitable and beneficial.

Innovation as a Catalyst for Progress

Despite the initial resistance, history demonstrates that innovation has consistently driven progress. The printing press democratized knowledge, fueling movements like the Renaissance and the Reformation. The industrial revolution, while disrupting traditional craftsmanship, paved the way for mass production, economic growth, and improved living standards. Today, digital technologies are transforming industries and creating opportunities unimaginable a century ago.

What these examples reveal is that the benefits of innovation often outweigh its risks once society adapts to the change. Tools that were once viewed with suspicion have become indispensable, reshaping how we live, work, and create. The printing press, once feared as a disruptor of sacred knowledge, became a cornerstone of education and cultural exchange. Mechanized looms, criticized for eroding craftsmanship, laid the groundwork for industries that provide livelihoods to millions. Even artificial intelligence, despite current debates, holds immense potential to enhance creativity, efficiency, and problem-solving across domains.

The Importance of Dialogue and Inclusivity

One of the key lessons from history is the importance of addressing the concerns of those impacted by new tools. Resistance often arises from genuine fears—about job displacement, loss of identity, or the ethical implications of technology. Dismissing these concerns as mere technophobia risks alienating those who feel threatened, deepening divides and hindering progress.

Instead, fostering dialogue and inclusivity is essential. Engaging with stakeholders, listening to their concerns, and involving them in shaping the adoption of new technologies can help alleviate fears and build trust. For example, providing training programs to help workers transition to new roles, creating safeguards against misuse, and ensuring that the benefits of innovation are widely distributed can make technological change more equitable and less disruptive.

Harnessing the Transformative Power of Creativity and Ingenuity

Ultimately, the historical pattern of skepticism followed by acceptance underscores the resilience and adaptability of the human spirit. While new tools may disrupt the status quo, they also unlock new possibilities for creativity and ingenuity. The key lies in recognizing that innovation is not an end in itself—it is a means to improve lives, solve problems, and expand the horizons of what is possible.

By learning from history, we can navigate the challenges of technological change with greater foresight and empathy. We can address resistance not as an obstacle but as an opportunity to refine and improve our approach to innovation. In doing so, we not only overcome the barriers to progress but also ensure that the transformative power of creativity and ingenuity benefits society as a whole.

The lesson is clear: skepticism is natural, but it is not insurmountable. With thoughtful dialogue, inclusive strategies, and a commitment to equitable progress, we can embrace new tools as catalysts for a better future.

Summary

This chapter explores how technology has historically intersected with and amplified human creativity. From the invention of the printing press, which democratized access to knowledge, to the advent of computers and the internet, each technological leap has reshaped creative processes and expanded possibilities.

The focus then shifts to the present-day integration of AI and advanced tools in creative fields. Examples include AI-powered design software, music composition tools, and virtual reality in storytelling, which augment human capabilities while challenging traditional notions of originality and authorship. The chapter also acknowledges concerns such as the loss of human touch and ethical dilemmas arising from AI-generated content.

Ultimately, this chapter highlights the transformative potential of technology in amplifying creativity while emphasizing the need to balance innovation with authenticity and ethical responsibility.

3: The Machine Age

The Machine Age revolutionized humanity's relationship with technology, introducing mechanized tools that amplified physical capabilities and transformed industries. Today, we stand on the cusp of a new era, defined not by machines that extend our hands but by those that mimic, and even challenge, the workings of our minds. The rise of artificial intelligence (AI) marks a pivotal moment in this evolution, as machines transition from executing predefined tasks to learning, reasoning, and making decisions. Unlike any tool before it, AI embodies both promise and peril, inspiring visions of a future where human potential is amplified and fears of disruption that could redefine the very fabric of society. This chapter delves into what sets AI apart, exploring its transformative nature and the dual role it plays as both an amplifier of human ingenuity and a disruptor of established norms.

3.1 What Makes AI Different?

Artificial intelligence (AI) stands apart from the traditional tools and machines of the past, marking a fundamental shift in how we interact with technology. Unlike earlier innovations designed to amplify human physical or cognitive efforts, AI introduces the concept of machines that can think, learn, and make decisions. This capability redefines the boundaries of what tools can do, blurring the line between human ingenuity and machine-driven innovation. The transformative nature of AI lies in its ability to not only execute tasks but also adapt, improve, and operate autonomously.

From Tools to Thinkers: How AI Transcends Traditional Machines

Traditional machines have long been invaluable in amplifying human capabilities. From the lever to advanced robotics, these tools function as extensions of human intent, executing predefined tasks with precision and efficiency. Their utility, however, is limited by their reliance on explicit programming or mechanical design. They perform exactly as directed, with no capacity for adaptation or independent thought. Artificial intelligence (AI) represents a fundamental shift in this paradigm, introducing machines that not only follow instructions but also process information, learn from data, and generate insights that go beyond their initial programming.

Beyond Execution: Machines as Collaborators

AI transforms machines from passive executors of human commands into active collaborators. Unlike traditional tools that are confined to repetitive, pre-programmed tasks, AI systems can interpret data, identify patterns, and generate solutions autonomously. This capability elevates their role from mere instruments to partners in innovation.

For example, in healthcare, traditional diagnostic machines provide data such as X-rays or lab results, requiring human doctors to interpret them. AI systems, on the other hand, can analyze complex medical datasets to detect subtle patterns that might escape human observation. They can predict diseases based on genetic profiles, recommend personalized treatments, and even monitor patient progress in real time. This ability to process vast amounts of information and draw actionable insights has redefined the boundaries of what machines can contribute to healthcare.

Creativity Redefined: AI in Artistic and Intellectual Domains

One of AI's most striking departures from traditional machines is its ability to engage in creative processes. In industries like art, music, and literature, AI algorithms are now capable of generating content that rivals human creativity. Tools like generative adversarial networks (GANs) create realistic artwork, while AI-powered platforms compose symphonies, write poetry, or produce scripts for films.

What sets AI apart is its ability to combine and reinterpret data in novel ways. For instance, AI systems trained on thousands of paintings can produce original artwork that blends the styles of different artists or invents entirely new ones. Similarly, natural language processing models can write compelling narratives, generating storylines that surprise and engage their human collaborators. This capacity to contribute creative ideas positions AI not as a replacement for human creators but as a co-creator, opening new dimensions in artistic and intellectual expression.

The Cognitive Leap: Intelligence in Action

AI's leap from execution to cognition is what truly distinguishes it from traditional machines. While an industrial robot can assemble a car with flawless precision, it cannot adapt to a new design or improve its methods without human intervention. AI systems, however, learn from experience and refine their algorithms over time. This learning ability allows them to evolve alongside changing conditions, solving problems that were not explicitly anticipated during their development.

For example, in finance, AI-driven algorithms analyze market trends, predict fluctuations, and make investment decisions in real time. These systems adapt to new data, adjusting their strategies dynamically to maximize returns. Unlike static models, they continuously improve their performance, becoming more effective with each iteration. This adaptability enables AI to operate in dynamic environments where traditional machines would struggle.

Applications Across Industries

The transformative potential of AI extends to a wide range of industries. In transportation, self-driving cars use AI to process real-time data from sensors, cameras, and maps to navigate safely and efficiently. In agriculture, AI systems analyze soil conditions, weather patterns, and crop health to optimize farming practices. Even in areas like customer service, AI-powered chatbots provide personalized interactions, understanding user queries and responding intelligently.

These applications demonstrate how AI transcends the limitations of traditional machines, enabling them to take on roles that require analysis, decision-making, and innovation. The shift from tools to thinkers has redefined the relationship between humans and technology, creating opportunities for collaboration that were previously unimaginable.

A Transformative Force

AI's ability to process information, adapt to new challenges, and contribute creatively positions it as a transformative force in the modern era. Unlike traditional machines that are confined to static roles, AI evolves, learns, and collaborates, making it a dynamic partner in solving complex problems and driving innovation. By bridging the gap between execution and cognition, AI has redefined what machines can do and expanded the possibilities for human progress.

The Learning Machine: AI's Ability to Evolve

One of the most transformative aspects of artificial intelligence (AI) is its capacity to learn and evolve. Unlike traditional tools that rely on fixed programming or manual updates to adapt, AI systems use processes such as machine learning (ML) and deep learning to refine their operations continuously. This ability to improve autonomously makes AI a dynamic and resilient technology, capable of responding to changing environments and solving increasingly complex problems.

How AI Learns: The Core of Machine Learning

Machine learning is a subset of AI that allows systems to identify patterns in data and make decisions without explicit instructions. It operates by training models on datasets, enabling the system to recognize correlations and predict outcomes. For example, supervised learning uses labeled data to teach the AI system how to classify or predict specific outputs, while unsupervised learning explores unlabeled data to identify hidden patterns and clusters. Reinforcement learning, another method, involves the AI learning through trial and error, receiving feedback in the form of rewards or penalties.

Deep learning, a more advanced form of machine learning, uses artificial neural networks inspired by the human brain to process vast amounts of data. These networks consist of layers of interconnected nodes that analyze data hierarchically, allowing the system to tackle highly complex tasks such as image recognition, natural language processing, and speech synthesis. This layered approach enables AI to learn not just simple associations but intricate relationships and abstractions.

Dynamic Adaptation: Continuous Improvement in Real-Time

AI's ability to evolve lies in its capacity to adapt dynamically based on new data. Unlike traditional software, which requires human developers to make updates or improvements, AI systems can modify their algorithms as they process additional information. This adaptability ensures that AI remains relevant and effective, even in rapidly changing environments.

For instance, recommendation systems like those used by streaming platforms (e.g., Netflix, Spotify) or online retailers (e.g., Amazon) continuously learn from user behavior. Every click, search, and purchase provides data that the AI system analyzes to refine its recommendations. Over time, these systems become increasingly accurate, delivering personalized suggestions that align more closely with individual preferences.

Similarly, in transportation, autonomous vehicles leverage machine learning to interpret real-time environmental data. They analyze inputs from sensors, cameras, and GPS systems to make split-second decisions, such as identifying obstacles, predicting pedestrian movements, and adjusting to traffic conditions. As these vehicles encounter more scenarios, their decision-making algorithms improve, enhancing safety and reliability.

The Role of Feedback in AI Evolution

Feedback loops are integral to AI's learning process. Positive feedback reinforces successful actions, while negative feedback highlights areas for improvement. This iterative cycle allows AI to refine its performance over time, much like how humans learn from experience. For example, in fraud detection systems, AI models are trained on historical transaction data to identify patterns indicative of fraudulent activity. As the system encounters new fraud cases, it incorporates these examples into its training, improving its ability to detect and prevent future incidents.

In healthcare, AI-powered diagnostic tools use feedback from medical professionals to enhance their accuracy. An AI system that initially flags potential anomalies in medical images can refine its analysis based on doctors' confirmations or corrections, becoming a more effective diagnostic assistant with each iteration.

Versatility Across Domains

The capacity for self-improvement makes AI a versatile technology applicable across diverse fields. In finance, AI models adapt to market trends, enabling more accurate predictions and automated trading strategies. In agriculture, AI systems analyze weather patterns, soil conditions, and crop health to optimize farming practices. Even in creative industries, AI evolves by learning from vast datasets of artistic or musical works, enabling it to generate original compositions or visual art that push creative boundaries.

This adaptability also equips AI to address challenges that are unpredictable or highly dynamic. For example, during natural disasters, AI systems can analyze satellite imagery, weather data, and on-the-ground reports to provide real-time insights for disaster response and resource allocation. Their ability to learn and adapt in such scenarios makes them invaluable tools for solving urgent and complex problems.

The Implications of a Learning Machine

AI's learning capability represents a fundamental shift in how we think about technology. It moves beyond the static nature of traditional tools, positioning AI as a living system that grows in sophistication and effectiveness over time. However, this dynamic nature also raises important considerations. The reliance on data for learning underscores the need for careful data management, including issues of privacy, bias, and security. Additionally, as AI systems evolve, ensuring their alignment with human values and goals becomes a critical priority.

Ultimately, the ability of AI to learn and evolve is what makes it such a powerful and transformative force. By continuously adapting to new challenges and opportunities, AI systems can operate at the cutting edge of innovation, reshaping industries and redefining what technology can achieve. This unique characteristic ensures that AI remains not just a tool of the present but a cornerstone of the future.

AI and Autonomy: When Machines Make Decisions

The rise of artificial intelligence (AI) has introduced a profound shift in the way machines function, offering them a level of autonomy that traditional tools could never achieve. Unlike conventional tools, which rely entirely on human oversight and direct input to perform their tasks, AI systems can process vast amounts of data, apply decision-making algorithms, and operate independently within defined parameters. This capability enables them to act in real-time, often faster and more accurately than humans, and to address situations without waiting for explicit instructions. However, this newfound autonomy also raises critical ethical and practical questions about accountability, control, and trust in machines.

The Mechanics of Autonomy: Decision-Making in Real-Time

AI systems derive their autonomy from their ability to analyze data, detect patterns, and predict outcomes using sophisticated algorithms. This allows them to make decisions on the fly, adapting to changing conditions and optimizing their actions to achieve desired outcomes. Unlike traditional tools, which are limited to executing predefined functions, AI can assess complex scenarios and select the most appropriate course of action without human intervention.

For instance, autonomous vehicles process inputs from a multitude of sensors, cameras, and navigation systems to make instantaneous decisions about speed, lane changes, and obstacle avoidance. Similarly, AI-powered drones used in agriculture can monitor crop health, identify areas needing attention, and even apply fertilizers or pesticides autonomously. These systems demonstrate the potential of AI to handle tasks that require rapid decision-making in dynamic environments, outperforming human capabilities in speed and precision.

Real-World Applications of Autonomous AI

The applications of AI autonomy span a wide range of industries, revolutionizing workflows and enhancing efficiency. In financial markets, algorithmic trading systems analyze market conditions, assess risks, and execute trades in milliseconds, reacting to changes far faster than human traders ever could. These systems can process massive volumes of data, identifying opportunities and mitigating risks with unparalleled precision.

In customer service, AI-powered chatbots equipped with natural language processing interact directly with users, understanding their queries, resolving issues, and even anticipating their needs. This level of autonomy reduces response times and enhances customer satisfaction, all without human intervention. Similarly, in logistics and supply chain management, autonomous systems optimize routes, manage inventory, and predict demand, improving operational efficiency and reducing costs.

Even in healthcare, autonomous AI systems are proving transformative. AI algorithms analyze medical images to identify potential abnormalities, often with accuracy comparable to human experts. Robotic surgical systems guided by AI perform intricate procedures with minimal invasiveness, improving patient outcomes and reducing recovery times. These examples illustrate how AI autonomy is reshaping industries by enabling machines to act independently and efficiently.

Challenges and Ethical Implications of Autonomous AI

While the autonomy of AI offers immense benefits, it also raises significant challenges and ethical concerns. One of the most pressing issues is accountability. When an AI system makes a decision, especially one with high stakes, who is responsible for the outcome? For example, if an autonomous vehicle is involved in an accident, is the fault with the manufacturer, the programmer, or the vehicle itself? This lack of clarity complicates the integration of autonomous systems into society, particularly in industries where safety and liability are critical.

Another concern is the potential for biases in decision-making. AI systems rely on data to learn and make decisions, and if that data is biased, the outcomes can be skewed. For instance, a hiring algorithm trained on historical data might inadvertently perpetuate gender or racial biases present in the original dataset. This highlights the importance of ensuring that the data used to train AI systems is representative, fair, and ethically sourced.

Moreover, the growing autonomy of AI raises questions about control. As machines become more capable of independent decision-making, ensuring that their actions align with human values and intentions becomes increasingly important. Mechanisms such as explainable AI (XAI), which aims to make AI decisions more transparent, are being developed to address this issue. However, striking the right balance between autonomy and control remains a complex challenge.

Autonomy as a Double-Edged Sword

AI's ability to operate autonomously is both a strength and a potential vulnerability. On one hand, it enables efficiency, scalability, and innovation across industries, allowing machines to handle tasks that would be impossible or impractical for humans. On the other hand, it introduces risks that require careful consideration and management, from ethical dilemmas to technical failures.

To integrate autonomous AI successfully, it is essential to establish robust frameworks for accountability, transparency, and regulation. Developers and policymakers must work together to address the challenges posed by autonomous systems while harnessing their transformative potential. By doing so, society can maximize the benefits of AI autonomy while minimizing its risks.

AI's autonomy marks a new chapter in the evolution of technology, shifting machines from passive tools to active decision-makers. This capability not only expands what machines can do but also challenges how humans define responsibility, trust, and collaboration in a world increasingly shaped by intelligent systems. As AI continues to evolve, understanding and managing its autonomy will be crucial to ensuring that it serves as a force for progress and positive change.

The Role of Data: The Lifeblood of Artificial Intelligence

Data is the cornerstone of artificial intelligence (AI), the essential resource that fuels its learning, decision-making, and adaptability. Unlike traditional machines, which depend on fixed programming or mechanical design, AI systems derive their power from the ability to analyze, process, and learn from vast amounts of data. This dynamic relationship with data enables AI to operate with precision, evolve over time, and tackle increasingly complex challenges. However, the central role of data also introduces significant challenges and responsibilities, highlighting the need for careful management and ethical considerations.

Data as the Foundation of AI Functionality

At its core, AI is a data-driven technology. Whether through machine learning, deep learning, or other AI techniques, these systems analyze data to identify patterns, establish correlations, and make predictions. The quality, quantity, and diversity of data are critical in determining the accuracy and effectiveness of an AI model. High-quality, well-curated datasets ensure that AI systems can make precise decisions, while diverse data enables them to generalize across a broad range of scenarios.

For instance, in image recognition, AI models trained on extensive and varied datasets can accurately identify objects, faces, and environments in real-world applications. Similarly, in natural language processing, large datasets containing diverse linguistic patterns allow AI to understand and generate human-like text. Without data, AI systems lack the raw material needed to function, making data the lifeblood of their operations.

The Transformative Impact of Data in Key Industries

The reliance on data has revolutionized industries by enabling AI to optimize processes, improve outcomes, and unlock new possibilities. In healthcare, for example, AI systems analyze medical images, patient histories, and genetic data to detect diseases early, recommend personalized treatments, and predict patient outcomes. This data-driven approach enhances diagnostic accuracy and empowers clinicians to make informed decisions.

In finance, AI leverages historical transaction data to detect fraudulent activity, forecast market trends, and personalize financial advice for customers. Predictive analytics, powered by data, allows businesses to anticipate risks and opportunities, improving decision-making and reducing uncertainty.

Transportation is another sector transformed by data-driven AI. Autonomous vehicles process real-time data from sensors, cameras, and GPS to navigate roads safely and efficiently. Traffic management systems use data to optimize traffic flow, reduce congestion, and enhance commuter experiences. These examples illustrate how data empowers AI to revolutionize industries, improving both efficiency and outcomes.

Challenges Associated with Data Dependency

While data is the backbone of AI, its centrality also presents significant challenges. One of the most pressing issues is **data privacy**. In an era of interconnected systems and widespread data collection, ensuring that sensitive information is handled securely and ethically is paramount. Mismanagement or unauthorized use of personal data can lead to privacy violations, eroding trust in AI technologies.

Another challenge is **data bias**. AI systems are only as unbiased as the data they are trained on. If datasets contain biases—whether due to historical inequalities, incomplete representation, or systemic errors—AI models can inadvertently perpetuate or amplify these biases. For example, biased hiring algorithms trained on historical data may discriminate against certain demographic groups, reinforcing patterns of exclusion.

Data security is another critical concern. With vast amounts of sensitive data being collected and processed, AI systems become attractive targets for cyberattacks. Unauthorized access to data can lead to breaches, identity theft, or manipulation of AI models, posing risks to individuals and organizations alike.

The Need for Ethical and Responsible Data Practices

Given the pivotal role of data in AI, ethical and responsible data management is essential. Organizations must prioritize transparency in how data is collected, processed, and used. Clear consent mechanisms, robust security protocols, and accountability frameworks are crucial to maintaining public trust.

Efforts to address bias should focus on curating diverse and representative datasets, continuously monitoring AI outputs, and incorporating fairness metrics into model evaluation. Additionally, fostering collaboration between policymakers, technologists, and ethicists can help create regulations that balance innovation with ethical considerations.

Emerging technologies like federated learning, which allows AI models to train on decentralized data without transferring it to a central repository, offer promising solutions to privacy and security challenges. Such innovations highlight the importance of aligning data practices with evolving societal expectations and technological capabilities.

Data as the Key to AI's Potential

Data's transformative power lies not only in enabling AI to function but also in shaping its potential. By unlocking insights, optimizing processes, and predicting outcomes, data-driven AI has the capacity to redefine industries and improve lives. However, the ethical and practical challenges associated with data dependency underscore the need for thoughtful management and responsible innovation.

As AI continues to evolve, the role of data will remain central to its success and impact. Ensuring that data is used responsibly, ethically, and inclusively is not just a technical challenge—it is a societal imperative. By recognizing data as both the lifeblood of AI and a shared resource, we can harness its potential to build a future that benefits everyone.

What Makes AI Revolutionary

Artificial Intelligence (AI) is not just an incremental technological advancement—it is a paradigm shift that redefines the role of machines in human life. Unlike previous tools, which required direct human input and were limited to static, predefined tasks, AI introduces machines that can think, learn, and act autonomously. This evolution bridges the gap between human ingenuity and machine efficiency, creating a new class of tools capable of operating in ways that were once exclusive to human intelligence. Its revolutionary nature lies not only in its capabilities but also in how it challenges fundamental notions of creativity, decision-making, and responsibility.

Autonomy: A Defining Leap

The ability of AI to act autonomously sets it apart from traditional machines. Autonomous systems can analyze data, draw conclusions, and make decisions without constant human oversight. This independence allows AI to handle tasks in real time, responding dynamically to changing conditions. For example, autonomous vehicles navigate roads by processing vast amounts of sensory data, adjusting to traffic conditions and potential hazards without human intervention. Similarly, AI-powered chatbots provide seamless customer service, understanding and resolving user queries in real time.

This autonomy enables machines to operate in environments where speed, precision, and adaptability are critical. However, it also shifts the dynamics of accountability and trust. When AI makes decisions independently, questions arise about who or what is ultimately responsible for the outcomes—an ethical and philosophical challenge unique to this technology.

Learning and Evolving: AI's Dynamic Nature

Unlike traditional machines, which remain static unless reprogrammed, AI systems possess the ability to learn and evolve. Through techniques such as machine learning and deep learning, AI analyzes data, identifies patterns, and refines its algorithms to improve performance over time. This dynamic nature makes AI a living system, capable of adapting to new challenges and opportunities.

For instance, recommendation engines used by streaming platforms and e-commerce sites continuously learn from user behavior, delivering increasingly personalized suggestions. In healthcare, AI systems improve diagnostic accuracy as they analyze more medical data, enhancing their ability to detect diseases early and recommend effective treatments. This ability to evolve ensures that AI remains relevant and effective in rapidly changing environments.

Expanding Creativity and Problem-Solving

AI has redefined the boundaries of creativity and problem-solving, enabling breakthroughs that were once thought impossible. In creative industries, AI systems generate original art, compose music, and write stories, often collaborating with human creators to produce innovative works. These tools extend the reach of human creativity, offering new ways to explore ideas and express emotions.

In problem-solving, AI excels at tackling complex challenges that require processing vast amounts of data. For example, climate modeling relies on AI to analyze environmental data, predict trends, and develop strategies for mitigating climate change. Similarly, in engineering and manufacturing, AI optimizes processes, reduces waste, and accelerates innovation, enabling industries to meet global demands more efficiently.

Redefining Decision-Making

AI's ability to process information at unprecedented speeds has transformed decision-making across sectors. By analyzing massive datasets, AI systems uncover insights that humans might overlook, supporting more informed and strategic choices. In finance, AI-driven algorithms predict market trends, manage risks, and execute trades with precision. In public health, AI models predict disease outbreaks and optimize resource allocation during emergencies, saving lives and improving outcomes.

However, the integration of AI into decision-making also raises critical ethical questions. As machines take on roles traditionally reserved for humans, such as hiring, medical diagnosis, or legal judgments, ensuring fairness, transparency, and accountability becomes essential. These challenges underscore the need for responsible AI development and deployment.

A Tool for Transformation, Not Replacement

While AI is a powerful tool, its true potential lies in collaboration with human intelligence rather than replacing it. By automating routine tasks, AI frees humans to focus on higher-level thinking, creativity, and innovation. For example, AI-assisted systems in architecture and design allow professionals to explore complex geometries and optimize construction processes, enhancing both creativity and efficiency.

This collaborative dynamic positions AI as an amplifier of human capabilities rather than a competitor. It challenges us to rethink how we define roles, skills, and the relationship between humans and machines in a rapidly evolving world.

Navigating AI's Transformative Potential

What makes AI revolutionary is not just its technical capabilities but its profound societal implications. It challenges us to reimagine how we create, decide, and interact with technology. Understanding AI's unique attributes is essential to harnessing its potential responsibly. This requires addressing ethical concerns, fostering inclusivity, and ensuring that AI serves the collective good.

By embracing AI as a transformative force, humanity has the opportunity to unlock unprecedented progress while navigating its challenges thoughtfully and effectively. AI is not merely a tool of the future—it is a partner in shaping it.

3.2 The Dual Nature of AI: Amplifier or Disruptor?

Artificial intelligence (AI) embodies a paradoxical role in modern society. It acts as both an amplifier of human potential and a disruptor of established norms and systems. While its capabilities promise breakthroughs in innovation, efficiency, and creativity, they also challenge traditional notions of work, ethics, and control. This dual nature of AI highlights its power to drive progress and its potential to create significant disruptions, necessitating a careful balance between opportunity and risk.

AI as a Catalyst for Human Potential

Artificial intelligence (AI) is redefining what individuals and organizations can achieve, acting as a powerful enabler that enhances human potential in unprecedented ways. By automating routine tasks, optimizing decision-making, and pushing creative and intellectual boundaries, AI empowers people to focus on higher-order thinking, problem-solving, and innovation. Its transformative impact spans multiple domains, from healthcare and education to the arts and sciences, showcasing its ability to elevate human capability rather than replace it.

Freeing Humans from Routine Tasks

One of AI's most visible contributions lies in its ability to automate repetitive and time-consuming tasks. This automation liberates professionals from mundane activities, allowing them to allocate their energy and expertise to more meaningful and complex challenges. For example, in business settings, AI-driven tools handle administrative tasks such as scheduling, data entry, and report generation, reducing workload and increasing efficiency. In logistics, AI systems optimize supply chains by predicting demand, managing inventory, and automating order fulfillment, streamlining operations while minimizing errors.

By taking over these routine functions, AI not only saves time but also reduces the mental strain associated with repetitive tasks. This shift enables individuals to focus on strategic thinking, creativity, and decision-making, unlocking greater productivity and job satisfaction.

Enhancing Decision-Making Across Sectors

AI's ability to process vast amounts of data with speed and precision has revolutionized decision-making in industries where accuracy and timeliness are critical. In healthcare, for instance, AI-powered diagnostic tools analyze medical images, patient histories, and genetic data to detect diseases at an early stage. These tools provide doctors with insights that might otherwise go unnoticed, enabling them to make faster, more informed decisions and improve patient outcomes.

In finance, AI algorithms analyze market trends, assess risks, and predict economic shifts, empowering businesses and investors to make data-driven decisions. Similarly, in agriculture, AI systems process environmental data to optimize planting schedules, water usage, and pest control strategies, helping farmers increase yields while conserving resources. Across these sectors, AI enhances decision-making by delivering insights that drive better outcomes and foster innovation.

Expanding Creative Horizons

AI is also revolutionizing creative industries, acting as a collaborator that extends the boundaries of human imagination. Artists, musicians, and writers are increasingly using AI to explore new techniques, generate ideas, and produce innovative works. For example, AI-powered tools like generative adversarial networks (GANs) create visually stunning artwork by blending styles or inventing entirely new ones. Similarly, musicians use AI to compose melodies, experiment with harmonies, and refine their compositions.

In film and media, AI streamlines processes such as scriptwriting, video editing, and special effects creation, enabling creators to bring their visions to life with greater precision and efficiency. Rather than replacing human creativity, AI amplifies it, providing artists with new tools to experiment, innovate, and push their disciplines forward.

Accelerating Scientific Discovery

The scientific community has embraced AI as a tool for accelerating research and discovery. In fields such as genomics, climate science, and materials engineering, AI analyzes massive datasets to uncover patterns, predict outcomes, and simulate experiments. For instance, in drug development, AI identifies potential compounds for treatment, drastically reducing the time and cost required for discovery. In climate science, AI models weather patterns and simulates environmental changes, helping researchers understand and address global challenges.

By automating data analysis and providing predictive insights, AI enables scientists to focus on hypothesis generation, experimentation, and interpretation. This partnership between human intellect and machine efficiency accelerates the pace of innovation and broadens the scope of what is scientifically possible.

Empowering Education and Personal Growth

In education, AI is transforming how students learn and educators teach. Personalized learning platforms use AI algorithms to adapt content, pacing, and teaching methods to individual students' needs, ensuring that each learner receives tailored support. These platforms identify strengths and weaknesses, providing targeted exercises and resources to help students master concepts more effectively.

For educators, AI tools streamline administrative tasks, such as grading and lesson planning, allowing them to dedicate more time to instruction and mentorship. By making education more accessible, engaging, and effective, AI empowers learners to reach their full potential and equips teachers with the tools to foster deeper connections and better outcomes.

AI as a Partner in Progress

The role of AI as a catalyst for human potential lies not in replacing people but in amplifying their abilities. By taking over routine tasks, enhancing decision-making, expanding creative possibilities, and accelerating discovery, AI empowers individuals and organizations to achieve more than ever before. This collaboration between human ingenuity and machine intelligence redefines the possibilities of progress, fostering innovation that benefits society as a whole.

As AI continues to evolve, its role as a transformative partner in unlocking human potential will only deepen, creating opportunities for growth, creativity, and achievement in every facet of life.

Challenges of Automation: The Fear of Job Displacement

AI-driven automation has brought unprecedented efficiency and innovation to industries worldwide, but it also raises significant concerns about its impact on the workforce. As machines and AI systems become more capable of performing tasks traditionally carried out by humans, the specter of job displacement looms large. This challenge, while not new in the history of technological progress, takes on a unique dimension with AI due to its ability to replace both manual and cognitive roles.

Sectors Most Affected by Automation

The effects of automation are most pronounced in industries reliant on repetitive, labor-intensive, or highly structured tasks. In manufacturing, robots equipped with AI streamline assembly lines, handling everything from welding to quality control with speed and precision. Logistics companies deploy AI-powered systems for inventory management, route optimization, and automated warehousing, reducing the need for human intervention. Similarly, in customer service, chatbots and virtual assistants handle routine queries, offering 24/7 support while reducing the demand for human agents.

These advancements improve efficiency and reduce costs but often come at the expense of traditional jobs. Workers in these sectors face the daunting prospect of being replaced by machines, intensifying fears about economic insecurity and widening inequality.

The Broader Scope of AI's Impact

Unlike previous waves of automation, which primarily affected manual labor, AI has begun to encroach on jobs requiring cognitive skills. For instance, in finance, AI algorithms analyze data, predict trends, and make investment decisions, reducing the need for human analysts. In journalism, AI tools generate news articles, summaries, and reports, performing tasks once reserved for skilled writers. Even in healthcare, AI systems assist in diagnostics and treatment planning, taking on roles traditionally held by highly trained professionals.

This broader reach of AI heightens concerns that no job is entirely immune to automation. As AI systems improve and expand their capabilities, the fear of widespread unemployment grows, particularly for workers who lack the skills to adapt to new roles.

The Fear of Widespread Unemployment

The fear of job displacement is not without basis. Automation has already reduced demand for certain roles, particularly in low-skill and routine jobs. For example, the rise of self-checkout systems in retail and automated kiosks in fast food has displaced cashiers and counter staff. Similarly, autonomous vehicles threaten the livelihoods of truck drivers, delivery personnel, and taxi operators.

These trends exacerbate concerns about inequality, as workers in vulnerable positions are often the hardest hit. Without intervention, automation could deepen socioeconomic divides, creating a world where high-skill workers reap the benefits of AI while others face unemployment or underemployment.

Lessons from History: Disruption and Opportunity

History, however, provides a more nuanced perspective on the impact of technological change. The industrial revolution, for example, disrupted traditional manual jobs but also gave rise to entirely new professions in engineering, design, and management. Similarly, the advent of computers displaced certain clerical roles while creating opportunities in IT, software development, and data analysis.

These examples suggest that while automation may eliminate some jobs, it also generates new opportunities by creating industries, roles, and markets that did not previously exist. The challenge lies in managing the transition to ensure that workers can adapt to these changes and benefit from the opportunities created by automation.

The Need for Reskilling and Upskilling

One of the most effective ways to address job displacement is through reskilling and upskilling initiatives. Workers need access to training programs that equip them with the skills required for emerging roles, such as programming, data analysis, and AI system management. Vocational training and lifelong learning initiatives can also help workers transition to new careers, ensuring that they remain competitive in an evolving job market.

Governments, educational institutions, and businesses must collaborate to create accessible, affordable, and effective training programs. For example, partnerships between industries and universities can develop curricula tailored to the needs of specific sectors, while government subsidies can make training more accessible to workers in vulnerable positions.

Equitable Sharing of Benefits

Beyond reskilling, policymakers and businesses must work to ensure that the benefits of AI-driven automation are shared equitably. This includes creating policies that support displaced workers, such as unemployment benefits, wage subsidies, and job placement services. Companies implementing automation should also consider investing in their workforce, offering retraining opportunities and transitioning employees into new roles within the organization.

At a broader level, exploring concepts like universal basic income (UBI) could help mitigate the economic impact of job displacement, providing a safety net for workers while they adapt to changes in the job market.

Building a Balanced Future

The fear of job displacement is a legitimate concern, but it is not an insurmountable challenge. By embracing proactive strategies—such as reskilling programs, supportive policies, and equitable distribution of benefits—society can navigate the challenges of automation while maximizing its potential. The goal should not be to resist technological progress but to ensure that it uplifts as many people as possible, creating a future where innovation benefits all.

Ethical Concerns: Bias, Privacy, and Control

The integration of artificial intelligence (AI) into society has brought with it transformative possibilities but also significant ethical challenges. As AI systems increasingly influence decisions across diverse fields—hiring, healthcare, finance, law enforcement, and beyond—concerns about bias, privacy, and control have taken center stage. These issues are not merely technical problems but deeply rooted ethical dilemmas that demand thoughtful and proactive solutions to ensure AI serves humanity fairly and responsibly.

Bias in AI: The Risk of Perpetuating Inequalities

AI systems are only as unbiased as the data they are trained on, and this dependency creates a significant vulnerability. If training data reflects historical biases, societal prejudices, or incomplete representation, AI systems can perpetuate or even amplify these issues in their outputs. For example, biased hiring algorithms trained on datasets that favor certain demographic groups may unintentionally disadvantage women, minorities, or older applicants, reinforcing systemic inequalities.

In criminal justice, predictive policing algorithms trained on biased crime data have been shown to disproportionately target certain communities, leading to unfair policing practices. Similarly, facial recognition technologies have demonstrated higher error rates for people with darker skin tones due to underrepresentation in training datasets.

The impact of bias extends beyond fairness—it undermines trust in AI systems and exacerbates existing societal divides. Addressing this issue requires a multifaceted approach, including ensuring diverse representation in training datasets, applying fairness metrics during model evaluation, and continuously auditing AI systems to identify and mitigate biases.

Privacy Concerns: Protecting Personal Information in the AI Era

AI's reliance on vast amounts of data raises critical questions about privacy. To function effectively, AI systems collect and process sensitive information, ranging from personal health records to browsing behaviors and location data. While this data enables AI to deliver personalized services and insights, it also creates significant risks if mishandled.

Data breaches and unauthorized access to personal information can have far-reaching consequences, including identity theft, financial loss, and reputational damage. For instance, healthcare AI systems processing patient data must ensure compliance with privacy laws like GDPR or HIPAA, as any lapses could erode public trust and compromise patient safety.

Additionally, the sheer volume of data collected raises ethical questions about consent and transparency. Are individuals fully aware of how their data is being used? Do they have control over whether or not their information is included in training datasets? These questions highlight the importance of clear consent mechanisms, robust data anonymization practices, and transparent data usage policies to protect individual privacy in an increasingly AI-driven world.

Control: The Challenge of Managing Autonomous AI Systems

As AI systems become more autonomous, the issue of control becomes increasingly complex. Autonomous AI can make decisions and take actions without human intervention, raising concerns about accountability and unintended consequences. For example, self-driving cars must make split-second decisions in life-and-death scenarios, prompting ethical dilemmas about whose safety to prioritize in unavoidable accidents.

Moreover, the potential misuse of AI systems for malicious purposes, such as deepfake technology used for misinformation or AI-powered cyberattacks, underscores the need for oversight. Ensuring that AI aligns with human values and intentions requires clear accountability mechanisms, such as explainable AI (XAI) frameworks that make AI decisions transparent and interpretable.

Control also extends to governance. Who should regulate AI, and how? The rapid pace of AI development often outstrips the ability of regulatory bodies to keep up, leading to gaps in oversight. This highlights the need for agile and adaptive regulatory frameworks that can evolve alongside technological advancements.

Transparency and Inclusivity: Foundations for Ethical AI

To address these ethical concerns, transparency and inclusivity must be at the forefront of AI design and deployment. Developers have a responsibility to ensure that AI systems are not only effective but also fair and accountable. This includes:

1. **Prioritizing Fairness:** Incorporating fairness metrics during the development process to identify and mitigate biases.
2. **Ensuring Diverse Representation:** Training AI models on datasets that reflect the diversity of the populations they serve, reducing the risk of exclusion or discrimination.
3. **Implementing Robust Privacy Protections:** Developing secure systems that safeguard personal information and give users control over how their data is used.

4. **Establishing Clear Accountability:** Creating mechanisms to trace and explain AI decisions, ensuring that responsibility can be assigned in case of errors or misuse.

Evolving Regulatory Frameworks

Regulation plays a critical role in addressing the ethical challenges of AI. Policymakers must work closely with technologists, ethicists, and industry leaders to develop comprehensive guidelines that balance innovation with responsibility. Key priorities for regulation include:

- **Standardizing Ethical Practices:** Defining universal ethical standards for AI design, deployment, and usage.
- **Ensuring Compliance:** Establishing oversight bodies to monitor adherence to ethical guidelines and penalize violations.
- **Promoting Collaboration:** Encouraging international cooperation to address global challenges, such as cross-border data sharing and AI governance.

A Path Forward: Responsible AI Deployment

Bias, privacy, and control are not insurmountable challenges but rather opportunities to shape AI in ways that reflect humanity's highest values. By proactively addressing these concerns, society can build trust in AI technologies and ensure they are used to uplift rather than harm. Ethical AI is not just a technical goal—it is a societal imperative that requires collaboration, vigilance, and a commitment to fairness and accountability. As AI continues to evolve, its success will depend on our ability to navigate these ethical complexities responsibly.

A Double-Edged Sword: Balancing Opportunity and Risk

Artificial intelligence (AI) holds a transformative power unparalleled in modern history. Its dual nature, as both an amplifier of human potential and a disruptor of societal norms, underscores the complexities of its integration into our lives. On one hand, AI promises revolutionary advancements, offering solutions to complex problems, enhancing productivity, and driving innovation. On the other, it presents significant challenges, including threats to employment, ethical dilemmas, and the risk of societal instability if not managed responsibly. Striking the right balance between these opportunities and risks is not merely a technical task but a societal imperative that requires thoughtful planning, collaboration, and vigilance.

The Unprecedented Opportunities of AI

AI's potential to address some of humanity's most pressing challenges is extraordinary. In healthcare, AI-powered diagnostics and personalized medicine are transforming patient care, enabling earlier detection of diseases and more effective treatments. In climate science, AI models predict weather patterns, optimize renewable energy usage, and aid in disaster response planning, helping mitigate the impacts of climate change.

In business and industry, AI drives efficiency by automating routine tasks, optimizing supply chains, and enhancing decision-making. Creative fields, too, are benefiting from AI as artists, writers, and designers collaborate with intelligent tools to push the boundaries of innovation. These applications illustrate AI's ability to amplify human ingenuity, open new frontiers, and create value across diverse domains.

The Risks of Disruption

However, the transformative nature of AI comes with significant risks. Automation driven by AI threatens to displace jobs across multiple sectors, particularly in roles involving repetitive tasks. While history suggests that technological shifts often create new opportunities, the pace of AI-driven change could leave many workers struggling to adapt, exacerbating economic inequality.

Ethical concerns also loom large. Issues such as data privacy, algorithmic bias, and the misuse of AI technologies challenge the responsible deployment of AI. Autonomous systems, capable of making decisions without human oversight, raise critical questions about accountability, especially in high-stakes scenarios such as self-driving vehicles or AI in law enforcement.

Societal stability is another concern. If the benefits of AI are not equitably distributed, it could deepen divides between those who can harness its potential and those who are left behind. The misuse of AI for malicious purposes, such as creating deepfakes or launching cyberattacks, further underscores its disruptive potential.

Proactive Planning and Collaboration

Navigating the dual nature of AI requires proactive measures that address both its opportunities and risks. Governments, businesses, and civil society must work together to establish frameworks that guide AI's development and deployment in ways that prioritize human well-being and fairness.

Education and Workforce Development: Investing in education and training programs is essential to prepare workers for the jobs of the future. Reskilling initiatives can help those displaced by automation transition into emerging roles in AI development, data analysis, and other high-demand fields. Lifelong learning programs can ensure that individuals remain adaptable in a rapidly changing job market.

Ethical Guidelines and Regulation: Establishing clear ethical standards and regulatory frameworks is critical to addressing concerns about bias, privacy, and accountability. Governments and international organizations must collaborate to create guidelines that balance innovation with protection, ensuring that AI serves the collective good.

Public Awareness and Inclusion: Engaging the public in conversations about AI's implications is vital to fostering trust and understanding. Transparent communication about how AI systems work, their benefits, and their risks can help demystify the technology and encourage informed participation in shaping its future.

Equitable Access to AI's Benefits: Efforts must be made to ensure that AI's advantages are shared broadly, reducing the risk of exacerbating existing inequalities. Policies that promote access to AI technologies, especially in underserved communities, can help bridge the gap between those who benefit from AI and those who risk being left behind.

Shaping AI's Role in Society

AI is neither inherently good nor bad—it is a tool that reflects the intentions and values of those who design and deploy it. The challenge lies in ensuring that AI is developed and used in ways that align with societal goals, ethical principles, and human values. This requires ongoing dialogue between stakeholders, from technologists and policymakers to educators and community leaders.

Balancing opportunity and risk means embracing AI's potential while remaining vigilant about its pitfalls. By addressing challenges head-on and fostering collaboration across sectors, society can harness AI's transformative power to drive progress, improve lives, and create a more equitable future.

A Future of Responsible Innovation

The dual nature of AI emphasizes both its promise and its perils. As we continue to integrate AI into our lives, understanding its potential and pitfalls is essential to ensuring that it serves humanity in meaningful and equitable ways. By proactively addressing risks and building frameworks for responsible innovation, we can shape a future where AI amplifies human potential while safeguarding against its disruptive effects. This careful balance will determine whether AI becomes a force for progress or a source of division, and it is a balance worth striving for.

Summary

This chapter delves into the evolving definition of creativity in an era shaped by AI and machine learning. It emphasizes that creativity is no longer confined to traditional arts but now encompasses problem-solving, innovation, and the ability to integrate human intuition with machine precision.

The chapter outlines how AI has become a collaborator in creative processes, providing tools for ideation, design, and execution. Examples include AI's role in generating visual art, music, and even literature, where it acts as a source of inspiration and efficiency. However, the chapter stresses that the essence of creativity lies in human input—empathy, intuition, and cultural context—which machines cannot replicate.

Key discussions focus on leveraging technology to augment rather than replace human creativity. It encourages creators to adapt, learn, and balance technological tools with personal expression, ensuring their work remains authentic and impactful. The chapter concludes by redefining creativity as a harmonious blend of human ingenuity and technological advancement.

Part II: The Paradox of Technology in Creativity

"Technology is both a mirror and a muse—reflecting our creative ambitions while challenging us to reimagine what is possible. Its paradox lies in its ability to amplify our potential while demanding that we stay rooted in our humanity."

— Sandeep Chavan

4: A New Frontier

Creativity has always been a defining trait of humanity, but the rise of advanced tools, particularly artificial intelligence (AI), is reshaping the boundaries of what we can imagine and create. This new frontier merges human ingenuity with machine intelligence, offering unprecedented opportunities to amplify creative expression. From art and music to writing and design, AI-assisted tools are not just instruments—they are collaborators, challenging traditional notions of creativity and expanding its possibilities. As we navigate this exciting era, it becomes clear that creativity with tools is not a replacement for human originality but a partnership that redefines the landscape of innovation. This chapter delves into the transformative power of AI-enhanced creativity, exploring its potential and real-world impact across diverse domains.

4.1 Enhancing Creativity with Machines

The emergence of advanced tools, especially artificial intelligence (AI), is transforming how creativity is perceived and practiced. Machines are no longer mere instruments to execute ideas—they have become active partners in the creative process. By bridging human imagination with machine intelligence, this collaboration expands the boundaries of innovation, allowing creators to explore uncharted territories in art, design, science, and more. This section delves into how machines enhance creativity, breaking barriers and forging a new path for human-machine co-creation.

The Symbiosis of Human Imagination and Machine Intelligence

The dynamic relationship between human imagination and machine intelligence is transforming the landscape of creativity. This partnership goes beyond mere functionality; it is a synergistic interplay that leverages the complementary strengths of humans and machines. While humans bring abstract thinking, emotional depth, and cultural awareness, machines contribute computational power, pattern recognition, and the ability to process immense datasets at incredible speeds. Together, this fusion opens new frontiers for innovation across disciplines.

Distinctive Strengths: Humans and Machines in Harmony

Humans possess qualities that are inherently difficult to replicate—empathy, intuition, and the ability to contextualize creativity within cultural, social, and historical dimensions. These attributes shape works of art, literature, and design that resonate emotionally and intellectually with audiences. For instance, a writer crafts narratives that evoke powerful emotions, or an artist uses abstract forms to challenge perspectives—both uniquely human capabilities.

Conversely, machines excel at analyzing large-scale data and identifying patterns that might elude human perception. AI tools can explore vast repositories of art, music, and literature to uncover hidden connections, generate fresh ideas, and suggest novel approaches. By offering insights that would otherwise remain undiscovered, machines serve as catalysts for human innovation, expanding creative horizons and enabling creators to work with greater precision and perspective.

AI as a Creative Partner: Practical Applications

The collaboration between humans and AI is already evident across fields like visual arts, music, literature, and architecture. In visual arts, AI-powered platforms like DeepArt and Runway ML can analyze thousands of artworks to generate unique visual styles or unexpected color combinations, sparking new ideas for artists. These tools serve as creative starting points, inspiring exploration rather than delivering final results.

In music, AI systems analyze harmonic structures, rhythmic trends, and melodic patterns across genres. These insights encourage musicians to blend traditional sounds with innovative techniques, creating compositions that push the boundaries of musical expression. Similarly, in architecture, AI models simulate environmental conditions and generate multiple design possibilities, helping architects craft spaces that balance functionality, aesthetics, and sustainability.

Expanding the Creative Frontier

The human-machine partnership does more than enhance existing creative practices; it challenges the limits of what can be achieved. By leveraging AI as a collaborator, creators are exploring uncharted territories, resurrecting incomplete works, and developing modern reinterpretations of classic ideas. For example, AI has been used to complete unfinished symphonies and literature, offering a bridge between past creativity and contemporary reinterpretation.

Moreover, this collaboration democratizes access to creative tools, enabling individuals from diverse backgrounds to experiment and produce high-quality work. Aspiring creators, regardless of their technical expertise, can now leverage AI platforms to refine their skills, discover unique styles, and expand their creative potential. This inclusivity ensures that creativity is no longer constrained by access to specialized training or resources.

Navigating Challenges and Opportunities

While the synergy between human imagination and AI holds immense promise, it also raises challenges that require careful navigation. Ensuring that human authenticity remains central to creative processes is a key concern as AI's capabilities grow. The risk of over-reliance on machine-generated outputs must be counterbalanced by a commitment to preserving the human voice and vision.

Creators must also cultivate the mindset to use AI as an enhancer rather than a replacement. This involves understanding the potential of AI tools, learning how to integrate them effectively, and ensuring that the end result reflects their original intent and creativity.

Redefining Creativity for a New Era

The interplay between human creativity and machine intelligence heralds a new paradigm for innovation and artistic expression. Far from diminishing the role of humans, this relationship amplifies it, enabling creators to delve deeper into their ideas, experiment more boldly, and achieve results previously thought unattainable.

By embracing this collaborative potential, we are not merely adopting new tools but expanding the very definition of creativity. The future of art, design, and problem-solving lies in this evolving symbiosis—where human ingenuity is enhanced, not overshadowed, by technological advancement.

This partnership is a testament to the transformative power of imagination when supported by the boundless capabilities of machine intelligence.

From Automation to Inspiration: How Tools Fuel Innovation

Throughout history, tools have extended human capabilities, streamlining tasks, improving accuracy, and freeing time for more complex pursuits. Initially, these tools were utilitarian, designed primarily to solve practical problems and increase efficiency. However, the advent of artificial intelligence (AI) has transformed the role of tools, turning them into active participants in the creative process. AI-driven tools now do more than execute instructions—they inspire new ideas, challenge traditional thinking, and encourage creators to push the boundaries of what is possible.

The Shift from Efficiency to Creativity

Traditional tools focused on reducing manual effort and enhancing productivity. By automating repetitive tasks, they enabled humans to concentrate on creativity and problem-solving. AI has taken this evolution further by moving beyond automation to actively contributing to the creative process. Unlike earlier tools, which required explicit human direction, AI can analyze vast amounts of data, identify patterns, and offer insights that spark innovation.

For instance, in writing, AI platforms like ChatGPT or Sudowrite go beyond correcting grammar or suggesting synonyms. They propose narrative developments, enrich character arcs, and refine prose, transforming writing into a collaborative dialogue where the creator and the tool co-create. This shift enables writers to explore unexpected avenues, elevating their storytelling to new heights.

AI as a Catalyst for Creativity

AI's strength lies in its ability to reinterpret and remix ideas, providing creators with fresh perspectives and new possibilities. In the visual arts, generative AI tools like DALL-E and MidJourney synthesize styles, motifs, and themes from extensive visual datasets, producing outputs that inspire artists to venture beyond their conventional methods. These tools do not replace the artist's creativity; instead, they serve as companions that ignite imagination and encourage experimentation.

In fields like architecture and product design, AI algorithms analyze diverse factors such as functionality, environmental considerations, and aesthetic appeal to propose innovative solutions. These machine-generated prototypes become the foundation for human designers to refine and adapt, creating results that balance practical needs with visionary design.

Accelerating Creative Iteration

One of AI's most transformative contributions is its ability to speed up the iterative process. By rapidly generating multiple versions of a design, narrative, or concept, AI enables creators to evaluate and refine their ideas at unprecedented speeds. This iterative capability fosters experimentation, allowing creators to explore a wide array of possibilities and arrive at optimal outcomes more efficiently.

In the fashion industry, for example, AI tools analyze trends, fabrics, and consumer preferences to generate diverse design options. Designers can review these outputs, select elements that resonate with their vision, and incorporate them into their collections. Similarly, in filmmaking, AI aids directors by suggesting visual effects, cuts, and transitions, helping them realize their creative vision more effectively.

Transforming Creative Workflows

AI is reshaping workflows across creative industries, fostering collaboration and interactivity. In music, platforms like Amper Music and AIVA create melodies and harmonies based on minimal input, enabling musicians to experiment with new genres and styles. These tools act as collaborators, complementing the artist's vision and expanding their creative palette.

In product development, AI simulates real-world conditions to test prototypes virtually, minimizing the need for physical trials and accelerating innovation cycles. This integration allows creators to focus on refining their ideas while AI handles time-intensive technical details, enhancing both efficiency and quality.

Pushing the Boundaries of Human Creativity

The transition from automation to inspiration signifies a profound shift in how tools enhance creativity. AI does more than assist—it challenges creators to think in novel ways, embrace unconventional approaches, and explore uncharted territories. Acting as both a mirror and a muse, AI reflects human creativity while pushing it forward.

This dynamic interaction highlights AI's potential to amplify human ingenuity. It enables creators to transcend limitations, fostering a new era where imagination and technology work in harmony. In this evolving landscape, tools are no longer passive instruments; they are active collaborators, shaping the future of innovation and artistic expression.

Breaking Barriers: Expanding Creative Possibilities with AI

Artificial intelligence (AI) has become a transformative force in creativity, breaking traditional boundaries and redefining the scope of what can be achieved. By leveraging its ability to process vast amounts of data, uncover hidden patterns, and generate innovative insights, AI empowers creators across disciplines to transcend limitations. From art and design to filmmaking and scientific research, AI expands creative possibilities, offering new realms of imagination and functionality.

Transcending Human Limitations with Data and Insights

AI's core strength lies in its capacity to analyze complex datasets and identify patterns that elude human perception. This capability allows it to act as a catalyst for creativity, offering novel solutions and perspectives that push beyond conventional approaches.

In scientific research, tools like AlphaFold have revolutionized protein structure prediction, solving problems that have challenged experts for decades. This breakthrough is transforming drug discovery and biotechnology, demonstrating how AI removes barriers once thought insurmountable.

Similarly, in creative fields, AI analyzes trends, styles, and preferences, inspiring new approaches and challenging creators to think outside established norms. By dynamically interacting with human ingenuity, AI fosters innovation on a scale previously unimaginable.

Sustainable and Innovative Design in Architecture

In architecture, AI has become an essential partner, enabling the creation of designs that are not only functional and aesthetic but also sustainable. By analyzing environmental factors such as climate, energy efficiency, and site-specific conditions, AI algorithms generate insights that help architects craft environmentally conscious and innovative structures.

For instance, AI can recommend materials, layouts, and construction methods that minimize waste and reduce carbon footprints. These insights empower architects to balance creative expression with sustainability, creating buildings that align with modern ecological demands.

Additionally, AI's ability to simulate real-world scenarios enables architects to test and refine their designs before construction, ensuring concepts are both practical and visionary. This process transcends traditional trial-and-error methods, allowing for more efficient and effective innovation.

Enhancing Storytelling in Filmmaking

AI is revolutionizing storytelling in filmmaking, offering tools that analyze audience preferences, emotional triggers, and cultural trends to enhance narrative impact. These tools provide filmmakers with data-driven insights to create stories that resonate deeply with diverse audiences.

For example, AI systems analyze box office performance and streaming trends to suggest themes, character arcs, or visual styles that engage target demographics. In post-production, AI-powered tools streamline tasks such as color grading, sound editing, and special effects, enabling filmmakers to focus on their creative vision.

Generative AI also plays an active role in conceptualizing scripts, designing characters, and building immersive virtual environments, expanding the possibilities of cinematic expression. This collaboration between human creativity and AI precision paves the way for groundbreaking storytelling.

Fostering Innovation Across Disciplines

AI's influence extends beyond the arts into fields like fashion, product design, and engineering. In fashion, AI analyzes trends and consumer behavior to inspire innovative collections that merge style with functionality. In product design, AI generates prototypes optimized for aesthetics, ergonomics, and manufacturing efficiency, accelerating development cycles and reducing time-to-market.

In engineering, AI-powered simulations test prototypes in virtual conditions, minimizing the need for costly physical trials. This approach accelerates innovation, allowing engineers to refine their designs and deliver transformative products more efficiently.

Redefining the Limits of Creativity

AI's ability to process, analyze, and generate ideas at an unprecedented scale is reshaping the creative landscape. By breaking through barriers that once constrained human ingenuity, AI enables creators to envision and realize possibilities previously thought unattainable. Challenges become opportunities, inspiring bold advancements across art, science, and technology.

AI's role is not to replace human creativity but to enhance it—providing tools and insights that amplify innovation. As creators embrace this collaboration, the boundaries of their disciplines continue to expand, forging a future where creativity is as limitless as the imagination itself.

The Collaborative Edge: Humans and Machines as Co-Creators

The emergence of artificial intelligence (AI) as a creative collaborator represents a transformative shift in the way humans approach innovation and artistry. Rather than replacing human creators, AI enhances their capabilities, offering unprecedented precision, speed, and insight. This synergy between human intuition and machine intelligence positions both as co-creators, unlocking possibilities that neither could achieve alone and reshaping the future of creativity across disciplines.

Synergy of Strengths: Human Intuition Meets Machine Precision

Human-machine collaboration thrives on the complementary strengths of each. Humans bring abstract thinking, emotional resonance, and the ability to contextualize creativity within cultural and societal frameworks. Machines contribute computational power, data-driven insights, and the ability to process immense datasets with remarkable accuracy. Together, they form a partnership that accelerates innovation while retaining the emotional and contextual depth unique to human creativity.

In industrial design, for example, AI generates numerous prototypes, simulates real-world scenarios, and evaluates performance under varying conditions. Designers and engineers interpret this data to refine ideas, adding aesthetic details and practical considerations that machines alone cannot conceptualize. This collaboration not only speeds up the design process but also ensures that final products are innovative, functional, and aligned with human needs.

AI in Fashion: Data-Driven Creativity

The fashion industry exemplifies how AI can enhance creative processes. By analyzing datasets that include consumer preferences, historical trends, and emerging styles, AI provides designers with actionable insights. Armed with this information, designers can craft collections that resonate deeply with audiences while staying ahead of market demands.

For instance, AI might suggest color schemes or fabric combinations that align with upcoming seasonal trends. Designers, in turn, interpret these suggestions, adding their personal touch and creative flair. This collaboration allows for experimentation and innovation without being hindered by logistical constraints, enabling designers to focus on the artistry of their work.

Enhancing Artistic and Scientific Endeavors

In the arts, AI serves as a tool for experimentation, inspiring creators to push boundaries. Visual artists use generative AI to craft abstract designs, blending traditional and digital mediums in ways that challenge conventional aesthetics. Similarly, musicians collaborate with AI to explore new harmonies, compose melodies, and experiment with genres that merge human emotion with algorithmic complexity.

In science and architecture, AI supports problem-solving by offering novel perspectives. Architects leverage AI to analyze structural possibilities and optimize designs for sustainability, while scientists use AI to model biological processes or develop innovative materials. These collaborations demonstrate how human ingenuity, combined with machine intelligence, can lead to groundbreaking achievements that were previously unattainable.

The Future of Human-Machine Collaboration

The collaborative potential of AI shifts the narrative from competition to partnership, allowing humans and machines to transcend traditional limitations. This dynamic redefines creativity by introducing new methods and ideas that neither party could achieve independently.

In gaming, for instance, AI generates complex environments and scenarios, enabling developers to create immersive worlds. Game designers then build on these foundations, adding narrative depth and emotional engagement to craft compelling experiences. This partnership not only accelerates development but also elevates the creative output, illustrating the power of collaboration.

Redefining Creativity in the Age of AI

AI's integration into creative processes amplifies human originality rather than diminishing it. By providing tools to deepen exploration, streamline iteration, and push creative boundaries, AI empowers creators to achieve new levels of innovation. The fusion of human imagination with machine intelligence heralds a new era where creativity knows no bounds.

This evolving partnership challenges traditional notions of creation, underscoring that the future of creativity lies in collaboration. By embracing this synergy, creators unlock new dimensions of expression, ensuring that AI acts as an amplifier of human potential rather than a replacement for it.

The relationship between human intuition and machine precision is more than a technological advance—it is a redefinition of how creativity is practiced and understood. Together, humans and machines are shaping a future where the boundaries of art, science, and design are continually expanding, driven by the limitless possibilities of imagination and innovation.

4.2 Real-World Examples of AI-Assisted Creativity

The impact of artificial intelligence (AI) on creativity is not theoretical—it is being realized across diverse domains, revolutionizing how we approach and practice innovation. By collaborating with humans, AI tools are transforming the way art, music, writing, design, architecture, and even scientific research are conceptualized and executed. These real-world examples showcase how AI-assisted creativity is breaking barriers and expanding possibilities.

Art in the Age of AI: Transforming Visual Media

The advent of artificial intelligence (AI) has ushered in a transformative era for visual art, redefining how creativity is conceived, executed, and appreciated. Generative AI tools like DeepArt, DALL-E, and Runway ML empower artists to transcend traditional boundaries, blending styles, reinterpreting existing works, and generating entirely new forms of artistic expression. These tools act as collaborators, expanding the horizons of creativity and enabling works that were once unimaginable.

Expanding the Canvas of Creativity

AI has revolutionized the artistic landscape by providing creators with tools to experiment with styles and techniques that challenge conventional norms. Generative systems analyze vast datasets of artistic works, uncovering patterns and relationships that inspire novel creations. For example, artists can combine the fluid brushstrokes of Impressionism with the fragmented forms of Cubism, producing unique fusions that honor tradition while pioneering new visual narratives.

This technological collaboration fosters a sense of freedom and curiosity in creators. By reducing the constraints of manual processes, AI accelerates iteration and exploration. The creative process evolves into a dynamic dialogue between human intuition and machine intelligence, resulting in works that are not only innovative but also emotionally resonant.

Challenging Traditional Notions of Authorship

AI's role in visual art has sparked a profound rethinking of authorship and originality. When AI-generated artworks emerge from human input and guidance, they raise questions about ownership and the nature of creativity. Does the credit belong to the machine that executes the visual or the human who provides the vision and direction?

This debate gained prominence in 2018 when an AI-generated painting, *Portrait of Edmond de Belamy*, sold for over $400,000 at Christie's. Such milestones highlight the growing acceptance of AI as a creative collaborator while prompting audiences to reconsider the essence of artistry. Rather than diminishing the artist's role, these discussions emphasize the evolving nature of creativity in the age of technology.

Reimagining and Enhancing Existing Art Forms

AI's capabilities extend beyond creating original works to reimagining and enriching existing art forms. Tools like DeepArt use neural networks to transform photographs into paintings that emulate the styles of masters like Van Gogh or Picasso. This fusion of old and new perspectives breathes fresh life into familiar subjects, evoking new emotional responses and interpretations.

AI is also playing a pivotal role in preserving cultural heritage. By analyzing damaged or incomplete artworks, AI can generate plausible reconstructions, offering a glimpse into history while safeguarding it for future generations. These applications underscore AI's potential to not only innovate but also sustain and celebrate the legacy of art.

Integrating Human Emotion with Machine Precision

Despite its computational roots, AI-driven art retains the emotional depth that defines human creativity. The artist's vision, intent, and sensibilities guide the process, ensuring that the machine's outputs resonate with audiences. This interplay of human emotion and machine precision produces works that are both technically refined and deeply meaningful.

Artists are using AI to address pressing social and environmental issues, crafting visual narratives that provoke thought and inspire action. By generating compelling imagery and exploring unique perspectives, AI helps creators articulate messages that might be challenging to express through traditional means.

Democratizing Visual Art

AI tools are democratizing the art world by making advanced creative resources accessible to a wider audience. User-friendly platforms like Canva and Adobe's Sensei-powered tools enable individuals with little formal training to produce professional-quality visual art. These technologies break down barriers to entry, fostering inclusivity and encouraging diverse voices to participate in creative expression.

This democratization enriches the scope of visual media, reflecting a broader array of perspectives and experiences. As more people engage with AI-driven tools, the art world becomes more dynamic, inclusive, and representative of global creativity.

A New Era of Visual Media

The integration of AI into visual art heralds a new chapter in the evolution of creativity. By enhancing traditional practices, challenging established norms, and expanding the possibilities of expression, AI has become more than a tool—it is a transformative partner in artistic innovation.

This symbiosis between human ingenuity and machine intelligence is reshaping the boundaries of visual media, ensuring that creativity remains vibrant, inclusive, and impactful. In this new era, art continues to evolve, driven by the collaboration of emotion, imagination, and technology.

Music Meets Algorithms: Composing with Artificial Intelligence

The intersection of artificial intelligence (AI) and music has opened up unprecedented avenues for creativity, collaboration, and innovation. Far from diminishing the role of human musicians, AI has become a dynamic partner in the creative process, offering tools that inspire, complement, and challenge traditional approaches. By leveraging vast datasets and generating compositions tailored to specific inputs, AI systems such as Amper Music, AIVA, and OpenAI's MuseNet enable musicians to venture into uncharted territories, blending human intuition with algorithmic ingenuity.

AI as a Creative Partner

AI-driven composition tools have redefined how musicians approach their craft, transforming AI from a passive assistant into an active collaborator. These systems analyze parameters such as mood, tempo, genre, and instrumentation to create original compositions that align with the user's creative vision. The results provide musicians with a foundation to personalize and refine, seamlessly integrating human artistry with machine precision.

For instance, AIVA, an AI composer, has produced classical works that echo the styles of legendary composers like Beethoven and Mozart. Such tools inspire contemporary artists to bridge classical traditions with modern trends, resulting in music that appeals to diverse audiences while pushing creative boundaries.

Revolutionizing Film Scoring

The integration of AI into film scoring has streamlined the creative process, offering composers a starting point for experimentation and iteration. AI tools generate preliminary soundtracks, enabling composers to explore a variety of themes and motifs before finalizing their compositions.

For example, AI-generated music can set the mood for a suspenseful thriller or enhance the emotional depth of a romantic drama. These compositions serve as a springboard, with human composers adding their artistic touch to craft soundtracks that are innovative, resonant, and tailored to the story. By handling time-consuming groundwork, AI allows composers to focus on the emotional and narrative aspects of their work.

Expanding Creative Horizons in Popular Music

In popular music, AI has emerged as a catalyst for experimentation and genre-blending. Artists collaborate with AI systems to generate melodies, harmonies, and backing tracks, using these as a basis for their creative explorations.

For instance, AI-generated beats and synth patterns challenge musicians to step beyond familiar techniques, fostering innovation and leading to distinctive compositions. Electronic producers, in particular, leverage AI to craft experimental soundscapes that redefine their genres. This partnership between artist and algorithm encourages creativity that is both cutting-edge and accessible.

Breaking Barriers in Music Education and Accessibility

AI is democratizing music creation, making it more accessible to those with limited formal training. Platforms like Amper Music offer intuitive interfaces that enable beginners to compose music by selecting a few parameters, fostering creativity across a broader demographic.

In education, AI-driven tools personalize the learning experience, analyzing a student's progress and tailoring exercises or compositions to their skill level. This adaptive approach enhances engagement and accelerates skill development, equipping students with the tools they need to express themselves musically.

Blurring the Lines Between Human and Machine-Generated Music

The collaboration between human musicians and AI is challenging traditional notions of authorship and originality. AI-generated compositions, guided by human input, often display a level of complexity and creativity that blurs the distinction between human and machine.

For example, some AI-generated pieces have been performed by renowned orchestras, with audiences unable to discern their origins. These instances highlight AI's potential to contribute meaningfully to the world of music while sparking debates about the evolving definition of creativity.

Driving Innovation Across Genres

AI's ability to analyze and synthesize diverse musical styles is driving innovation across genres. Jazz musicians use AI to explore improvisation techniques, while electronic producers experiment with AI-generated rhythms and textures to push their soundscapes into new dimensions. By bridging traditional and contemporary influences, AI helps musicians create works that are both experimental and grounded in their artistic vision.

A Harmonious Future of Collaboration

The integration of AI into music composition heralds a new chapter in the evolution of musical creativity. By acting as a collaborator, AI enhances the creative process while inspiring musicians to explore new possibilities. This partnership underscores that the essence of music lies in its ability to convey emotions, stories, and ideas, regardless of the tools employed.

As AI continues to evolve, its role in shaping the future of music will grow, leading to compositions that blend human creativity with machine intelligence. This harmonious collaboration ensures that music remains a dynamic, transformative, and deeply human art form in the age of technology.

AI in Writing: From Co-Authors to Storytellers

The written word has long been a cornerstone of human expression, and artificial intelligence (AI) is now transforming how stories are imagined, structured, and shared. Acting as co-authors, AI tools empower writers to elevate their storytelling, explore new formats, and overcome creative challenges like writer's block. From journalism to fiction and content creation, AI is redefining the writing process, accelerating creativity, and expanding the narrative possibilities available to modern writers.

AI as a Creative Partner for Storytelling

AI tools such as ChatGPT, Sudowrite, and Jasper have become indispensable for writers seeking to enhance their work. These systems generate story ideas, suggest plot twists, and refine prose, enabling authors to maintain momentum and expand their creative boundaries.

For example, an author stuck on a plot twist might use AI to propose unexpected developments or explore alternative scenarios. Similarly, when crafting dialogue, AI can generate natural-sounding exchanges, allowing authors to focus on the broader emotional and thematic arcs of their stories. By assisting with such elements, AI encourages experimentation and innovation, helping writers think beyond their usual patterns.

Enhancing World-Building and Character Development

In fiction, world-building and character development are often among the most intricate aspects of storytelling. AI tools excel in generating detailed descriptions of settings, societal structures, and mythologies, giving writers a foundation to refine and adapt.

For example, an author developing a fantasy novel can use AI to create richly imagined worlds, complete with unique cultures and histories. Similarly, in character development, AI suggests personality traits, backstories, and realistic motivations, helping authors craft complex and relatable characters. This collaboration enriches narratives by deepening the emotional and contextual layers that drive the story.

Revolutionizing Journalism and Content Creation

AI has become a game-changer in journalism, streamlining the reporting process and enhancing productivity. From analyzing datasets to drafting initial reports and suggesting headlines, AI-powered tools allow journalists to focus on refining their stories for accuracy and depth.

For instance, AI can transcribe interviews, summarize research findings, or generate data-driven insights, freeing up time for investigative work. Similarly, in content creation for blogs, marketing, and social media, platforms like Jasper and Writesonic enable businesses and creators to produce engaging, high-quality content at scale, tailored to specific audiences.

Overcoming Writer's Block and Enhancing Productivity

Writer's block, a common challenge for authors, is another area where AI proves invaluable. By generating prompts, brainstorming ideas, or drafting initial paragraphs, AI helps writers overcome creative hurdles and maintain momentum.

For example, a novelist grappling with a challenging chapter might use AI to suggest how the scene unfolds, offering multiple possibilities that spark new ideas. The flexibility of AI-generated suggestions allows authors to experiment with tone, pacing, and perspective, opening new creative pathways and fostering productivity.

Challenging Traditional Writing Practices

The integration of AI into writing raises thought-provoking questions about authorship, creativity, and originality. As co-authors, AI tools challenge the notion that writing is solely a human endeavor, offering a collaborative dynamic that some purists resist but many embrace as an opportunity for innovation.

For instance, experimental narratives that blend human and AI-generated elements push the boundaries of traditional storytelling, creating works that reflect both human intuition and algorithmic input. This fusion demonstrates the potential for AI to actively contribute to the evolution of writing, reshaping how stories are conceived and experienced.

Exploring New Formats and Genres

AI's adaptability extends beyond traditional writing, enabling exploration into interactive storytelling and emerging genres. In gaming, for example, AI generates dynamic story arcs and dialogue that adapt to player choices, creating immersive, real-time experiences.

Similarly, in screenwriting, AI helps scriptwriters refine scenes, develop character arcs, and visualize narratives, enhancing the creative process for film and television. These applications illustrate AI's versatility as a creative partner, allowing writers to engage with formats that demand innovation and interactivity.

A Redefined Writing Process

The integration of AI into writing does not replace human creativity but amplifies it. Acting as a collaborator, AI challenges writers to rethink their craft, fostering innovation and pushing the boundaries of storytelling. This partnership results in narratives that are richer, more engaging, and reflective of the possibilities that arise from human-machine collaboration.

As AI continues to evolve, its role in storytelling will expand, offering writers new tools to express their ideas and connect with audiences. This transformation ensures that the art of writing remains dynamic, inclusive, and adaptable in the age of technology.

Design and Architecture: Pioneering New Aesthetics with AI

Artificial intelligence (AI) is transforming design and architecture by seamlessly merging creativity with functionality and efficiency. By processing vast datasets—spanning environmental factors, user preferences, and aesthetic trends—AI empowers designers and architects to generate innovative concepts that challenge traditional methods. This fusion of human ingenuity and advanced technology not only accelerates the creative process but also unlocks new possibilities, ushering in a new era of sustainable and visionary design.

Transforming Architectural Design

AI has redefined modern architecture, revolutionizing how spaces are imagined and brought to life. Tools like Grasshopper and Autodesk's generative design software enable architects to explore countless design iterations, optimizing structures for aesthetics, functionality, and sustainability. These programs consider variables such as energy efficiency, structural integrity, and visual appeal, delivering holistic design solutions.

For example, AI-driven tools have played a pivotal role in creating eco-friendly buildings that harmonize sustainability with artistic vision. By analyzing environmental data—such as solar exposure, wind patterns, and temperature fluctuations—AI recommends layouts, materials, and orientations that maximize energy efficiency. Architects refine these insights, ensuring the final design reflects both practical needs and creative aspirations.

Innovating Product and Furniture Design

AI is driving innovation in product and furniture design by enabling the creation of ergonomic, user-centric solutions. Designers leverage AI to analyze human behavior, body mechanics, and material properties, resulting in products that prioritize comfort, usability, and sustainability.

For instance, AI tools have been used to design chairs that adapt to an individual's posture or lighting systems that respond to ambient conditions. By simulating user interactions, AI empowers designers to refine their concepts for optimal functionality and aesthetic appeal. This approach enhances the user experience and expands the boundaries of design innovation.

Enhancing Urban Planning and Public Spaces

The influence of AI extends to urban planning, where it facilitates the conceptualization of public spaces that cater to diverse community needs. AI tools analyze population density, traffic flow, and environmental impact to provide planners with data-driven insights for designing smarter, more livable cities.

For example, AI has been used to design pedestrian-friendly streetscapes that prioritize walkability, accessibility, and green spaces. These insights, combined with human creativity, enable urban planners to craft environments that harmonize with natural ecosystems while enhancing the quality of life for residents.

Accelerating the Design Process

One of AI's most transformative contributions is its ability to accelerate workflows. Traditional design processes often involve extensive trial and error, requiring significant time and resources to refine ideas. AI dramatically reduces this burden by generating multiple design iterations in a fraction of the time, enabling creators to experiment and innovate more freely.

Generative design software, for example, can produce hundreds of variations for a building's layout or a product's shape based on specific parameters. Designers can select the most promising options and tailor them to meet their objectives, streamlining the process without compromising quality.

Pushing Aesthetic Boundaries

AI inspires designers and architects to explore aesthetic possibilities that might be difficult to conceive through conventional methods. Generative algorithms produce intricate, organic shapes and patterns, often drawing inspiration from nature. This capability enables the creation of structures that are both visually stunning and structurally sound.

For instance, AI has been used to design facades with intricate patterns that optimize light and ventilation or bridges with biomimetic forms that blend seamlessly into their surroundings. These innovations encourage creators to move beyond established design principles, fostering a new visual language that bridges art and technology.

Integrating Sustainability into Design

Sustainability has become a cornerstone of modern design and architecture, and AI plays a crucial role in meeting this challenge. By evaluating the environmental impact of materials, construction methods, and energy use, AI guides creators toward more sustainable practices without compromising functionality or beauty.

For example, AI models help architects identify renewable materials, minimize construction waste, and optimize energy use throughout a building's lifecycle. These advancements allow designers to create spaces that respect the environment while setting new benchmarks for sustainable innovation.

A Harmonious Future of Design and Architecture

The partnership between AI and human creativity is revolutionizing the fields of design and architecture. By blending data-driven insights with artistic intuition, this collaboration enables the creation of spaces and objects that are innovative, efficient, and inspiring. From sustainable buildings and ergonomic products to vibrant urban spaces and futuristic structures, AI-assisted design transcends problem-solving, imagining possibilities that redefine the essence of creativity.

As AI continues to evolve, its role in design and architecture will deepen, fostering a future where technology and artistry work hand in hand to shape creations that reflect humanity's highest aspirations. This partnership ensures that design remains a dynamic, inclusive, and transformative force in the age of technology.

Innovation Beyond Art: AI's Role in Scientific and Industrial Creativity

Artificial intelligence (AI) is widely celebrated for its revolutionary impact on art and creativity, but its influence extends well beyond aesthetics. In scientific research and industrial design, AI is redefining innovation by analyzing vast datasets, uncovering hidden patterns, and simulating complex processes. It has become an indispensable tool for addressing challenges, accelerating breakthroughs, and expanding the boundaries of human knowledge and ingenuity. From revolutionizing medicine to optimizing engineering workflows, AI is transforming the way we approach scientific and industrial creativity.

AI in Scientific Research: Accelerating Discovery

AI is reshaping scientific research by streamlining processes that were traditionally time-intensive and resource-heavy. Its ability to process, analyze, and synthesize massive datasets has made it an invaluable ally across diverse fields, including healthcare, environmental science, and astronomy.

In healthcare, AI has revolutionized drug discovery by rapidly identifying potential compounds for treatment. Tools like DeepMind's AlphaFold predict protein structures with extraordinary accuracy, enabling researchers to understand diseases more effectively and design targeted therapies. What once took years can now be accomplished in months, dramatically accelerating progress in medical research.

Beyond healthcare, AI contributes significantly to environmental science by analyzing climate data to predict weather patterns, assess the impact of deforestation, and optimize renewable energy systems. In astronomy, AI processes vast amounts of telescope data to uncover celestial phenomena such as exoplanets and supernovae that might otherwise go undetected. These advancements underscore AI's ability to drive discovery across disciplines, making the impossible achievable.

Industrial Design: Pioneering Efficiency and Innovation

In the realm of industrial design, AI empowers engineers and manufacturers to create products that are efficient, innovative, and sustainable. By simulating real-world conditions and testing multiple design iterations, AI accelerates the development process while minimizing costs and resource consumption.

For example, in the automotive industry, AI assists in designing vehicles with optimized aerodynamics and fuel efficiency. By analyzing airflow, material properties, and energy consumption, AI generates prototypes that balance performance with sustainability. Similarly, electric vehicle manufacturers use AI to design advanced battery systems that enhance energy storage while reducing weight and cost.

In aerospace engineering, AI plays a critical role in developing lightweight, durable components for aircraft. Generative design tools explore thousands of configurations to meet stringent performance and safety standards. These innovations also extend to space exploration, where AI contributes to cutting-edge technologies like reusable rockets and autonomous spacecraft, pushing the boundaries of human ingenuity.

Enhancing Manufacturing Processes

AI is revolutionizing manufacturing by optimizing production lines, reducing waste, and predicting maintenance needs. Smart factories equipped with AI-powered sensors and robotics adapt to real-time conditions, ensuring seamless operations and minimizing downtime. Predictive maintenance, driven by AI's ability to analyze machine data, identifies potential issues before they escalate, reducing costs and disruptions.

Quality control has also been transformed by AI, with computer vision systems detecting product defects more accurately than human inspectors. This precision ensures consistent quality while minimizing waste, contributing to more sustainable manufacturing practices.

Bridging Creativity and Functionality

AI's role in scientific and industrial creativity goes beyond efficiency—it enables experimentation with innovative solutions previously out of reach. Architects and engineers, for example, use AI to design structures capable of withstanding extreme conditions such as earthquakes or hurricanes, all while maintaining aesthetic appeal.

In material science, AI accelerates the discovery of compounds with unique properties, such as ultra-strong or highly conductive materials. These breakthroughs have applications across industries, from electronics to construction, demonstrating AI's capacity to merge creativity with functionality to solve complex challenges.

Collaborative Innovation Across Disciplines

One of AI's most transformative contributions is its ability to foster interdisciplinary collaboration. By bringing together expertise from multiple fields, AI facilitates the development of holistic solutions that address multifaceted problems.

For instance, in the creation of wearable medical devices, AI integrates insights from healthcare, engineering, and design to develop products that are both effective and user-friendly. Similarly, AI-powered virtual simulations are transforming education and training, offering professionals hands-on experience in safe environments. From piloting aircraft to performing surgeries, these simulations reduce risks while enhancing skills.

A Catalyst for the Future

AI's transformative role in scientific and industrial creativity highlights its versatility as a tool for progress. By enabling faster iteration, reducing resource consumption, and fostering innovation, AI empowers researchers, engineers, and designers to address some of humanity's most pressing challenges. Its potential to reshape industries and redefine the boundaries of knowledge ensures that AI's impact will continue to grow.

As we move forward, the partnership between human ingenuity and AI-driven insights promises to drive groundbreaking discoveries and innovations. This evolving relationship underscores the profound potential of AI—not just as a tool for efficiency but as a source of inspiration and creativity that transcends the limits of traditional artistic expression.

A New Era of Creativity

The rise of AI-assisted creativity marks a pivotal moment in human history, redefining the boundaries of imagination and innovation across disciplines. From the vivid landscapes of digital art and algorithmic music to groundbreaking scientific discoveries and precision-driven industrial designs, AI is transforming how we approach creation. By fostering a partnership between human intuition and machine intelligence, AI is expanding the realm of possibilities, challenging traditional norms, and reshaping what creativity can achieve.

Blurring the Lines Between Human and Machine Creativity

At the core of this transformation is the evolving relationship between human creators and AI. Unlike traditional tools that assist with execution, AI contributes actively to the creative process, analyzing data, generating insights, and offering fresh perspectives. This collaborative dynamic has blurred the lines between human and machine creativity, resulting in outputs that combine emotional depth with computational precision.

Artists use generative AI to explore uncharted styles and techniques, unlocking visual expressions that may have remained undiscovered. Musicians collaborate with algorithms to compose pieces that intertwine human emotion with algorithmic complexity. This interplay not only enhances the quality of creative output but also challenges creators to rethink their processes, pushing the boundaries of their imagination.

Redefining the Creative Landscape Across Disciplines

AI's influence extends across diverse fields, each benefiting uniquely from its capabilities. In the visual arts, AI enables immersive digital experiences and interactive installations, captivating audiences in ways traditional media cannot. In literature, AI tools assist authors by generating intricate plots, refining prose, and experimenting with novel formats that stretch the boundaries of narrative structure.

In science and engineering, AI accelerates problem-solving and innovation. Researchers use AI to model biological systems, predict material behaviors, and design products that balance aesthetics, functionality, and sustainability. Architects and engineers employ AI to create structures that are not only visually striking but also environmentally conscious.

This cross-disciplinary reach highlights AI's ability to transcend enhancement—it fundamentally transforms practices by encouraging experimentation and fostering a culture of innovation that bridges fields and disciplines.

The Endless Potential of Human-Machine Collaboration

The synergy between human creativity and AI-driven insights has become a powerful catalyst for endless possibilities. By handling complex calculations, analyzing vast datasets, and generating novel ideas, AI amplifies human potential. This allows creators to focus on refining concepts, infusing them with personal meaning, and elevating their work to new heights.

The collaborative potential of AI is particularly profound in education and healthcare. In education, AI designs personalized learning experiences, adapting to the unique needs of each student and fostering deeper engagement. In healthcare, AI assists in creating more accurate diagnostic tools and tailored treatment plans, seamlessly blending technological precision with human compassion.

A Future Without Boundaries

As AI continues to evolve, its ability to expand creative horizons will only grow. Emerging advancements such as quantum computing, sophisticated neural networks, and next-generation generative algorithms promise to enhance AI's capabilities, enabling it to address challenges and opportunities that are currently beyond reach.

The future of creativity lies in embracing this partnership between human ingenuity and machine intelligence. Together, they will explore uncharted territories, from crafting immersive virtual realities to designing solutions for global challenges and composing symphonies that transcend cultural boundaries. The possibilities are limitless, bound only by the scope of our collective imagination.

A Testament to Human Innovation

The rise of AI-assisted creativity is more than a technological milestone—it is a testament to humanity's relentless pursuit of progress. By developing tools that amplify our abilities, we reaffirm the power of imagination and the importance of collaboration. Far from diminishing human originality, AI has the potential to elevate it, ensuring that creativity remains vibrant in an increasingly complex world.

As we integrate AI into creative practices, our understanding of what it means to create will continue to evolve. This partnership invites us to reimagine the possibilities of creativity, embracing a future where the boundaries of imagination truly know no limits. This new era of creativity is not just a step forward—it is a profound redefinition of how humanity shapes its world, fueled by the synergy of human and machine working together to inspire and innovate.

Summary

This chapter explores the dynamic partnership between humans and machines, emphasizing AI's role as a collaborator rather than a replacement in the creative process. It highlights how AI enhances human creativity by providing computational power, data insights, and innovative possibilities, while humans bring emotional intelligence, cultural sensitivity, and originality.

The chapter provides sector-specific examples of this collaboration, such as:

✓ **Design and Visual Arts**: AI tools generating design variations, which artists refine to align with their creative vision.

✓ **Storytelling**: AI suggesting character arcs or themes, leaving the final narrative infused with the author's personal experiences.

✓ **Scientific Research**: AI accelerating data analysis, enabling researchers to focus on hypothesis development and interpretation.

It also underscores the importance of maintaining human oversight to ensure AI's outputs align with ethical standards and creative intentions. Transparency in the use of AI, coupled with a balanced approach, fosters trust and innovation.

The chapter concludes by envisioning a future where human-AI collaboration unlocks unprecedented opportunities for creativity, enabling creators to push boundaries while staying true to their unique voices.

5: Creativity without Tools

In an age where tools and technology dominate creative processes, there remains a steadfast belief in the power of raw human ingenuity. For many purists, creativity is not about the tools used but about the unfiltered expression of ideas, emotions, and imagination. They argue that relying on tools can dilute the authenticity of creation, favoring simplicity and personal effort over technological enhancement. This chapter delves into the purist's perspective, exploring the value of creating without tools and examining the stories of those who reject technological aids to preserve the essence of unadulterated creativity.

5.1 The Value of Raw Human Ingenuity

The debate between tool-assisted and raw creativity is not new. At its core lies the belief that the essence of creativity stems from the human mind itself, unfiltered by external intervention. Purists argue that true creativity emerges not from the efficiency of tools but from the authenticity of effort, intuition, and emotion. This section explores the profound value of raw human ingenuity, shedding light on its unique ability to inspire, adapt, and innovate.

The Essence of Creativity: Inspiration Without Intervention

Creativity, at its purest, is often regarded as a deeply personal and intuitive process that transcends tools and external influences. For purists, the journey of creation—marked by introspection, trial, and discovery—holds as much importance as the final result. This philosophy emphasizes an unbroken connection between the creator and their work, one that is unencumbered by mediating tools or technologies.

At its heart, this perspective frames creativity as an intimate dialogue between the mind and the medium, fueled by inspiration and shaped by the creator's singular vision. Tools, while undeniably useful, are seen as potential disruptors—capable of imposing biases, diluting the creative process, or overshadowing the raw emotional essence of the work. Under this view, the act of creation becomes a spiritual endeavor, where the purity of human thought and effort is paramount.

Historical Masterpieces: Unassisted Creation at Its Best

History offers innumerable examples of masterpieces born from the direct, unmediated interaction between creators and their chosen mediums. These works are celebrated not only for their authenticity but also for the emotional depth and raw ingenuity they embody, untouched by the complexities of external tools.

Imagine the sculptor chiseling marble by hand, relying solely on their vision and the physical resistance of the stone. The resulting statue stands as a testament to their skill, perseverance, and intimate connection with their craft. Similarly, a poet crafting verses by candlelight captures fleeting moments of inspiration unfiltered by distractions or external aids. These creations, born from unassisted processes, amplify the focus and intention of the creator, fostering an environment where creativity flourishes uninhibited.

The Intimacy of Direct Creation

Creating without tools fosters a rare and profound intimacy between the creator and their work. Every stroke of a brush, every word written, or every note played becomes an unmediated extension of the creator's emotions and thoughts. This directness ensures that the final product reflects the maker's unique perspective and personal touch.

Consider a musician composing a melody on a simple acoustic instrument. Without digital effects or recording technology, the process is stripped down to its essence—where the music is shaped entirely by the musician's intent and skill. This unadulterated expression is what purists argue makes such creations deeply resonant and enduring.

The Double-Edged Sword of Tools

While tools can enhance efficiency and expand creative possibilities, purists often view them as a double-edged sword. Tools, by their very nature, introduce constraints—preset functionalities, algorithmic biases, or technical limitations—that can subtly steer creativity in unintended directions.

This concern has grown in an era dominated by advanced technology, where AI-generated content and automated tools often blur the line between human creativity and machine intervention. Purists argue that an over-reliance on these tools risks reducing creativity to a mechanical process, stripping away the personal introspection and emotional depth that define truly meaningful work.

Why Purity in Creativity Still Matters

The essence of creativity lies in its ability to reflect the human spirit—its emotions, struggles, and triumphs. By embracing an unassisted approach, creators reconnect with this essence, ensuring that their work embodies their vision and individuality. It is this purity that gives such creations their timeless appeal, resonating with audiences across generations.

In a world increasingly driven by convenience and efficiency, the act of creating without tools serves as a powerful reminder of the value of effort, perseverance, and personal touch. It elevates creativity from a functional activity to an act of self-expression and discovery, reaffirming its role as a uniquely human pursuit.

This philosophy of unassisted creativity is not a rejection of progress but a celebration of the raw potential of human ingenuity. It reminds us that, at its core, creativity is about connection—between the creator and their work, their audience, and ultimately, themselves. This unmediated connection transforms creativity into more than a skill; it becomes a profoundly enriching and timeless experience.

The Role of Intuition and Emotion in Creation

At the heart of raw creativity are intuition and emotion—uniquely human qualities that breathe life and meaning into the act of creation. These elements imbue creative works with authenticity, allowing them to resonate deeply with audiences in ways that tools, confined by logic and parameters, cannot replicate. Intuition and emotion are not merely components of creativity; they are its essence, guiding creators to craft profoundly personal yet universally impactful pieces.

Intuition: A Compass for Creativity

Intuition acts as a creator's inner compass, silently guiding decisions that may defy conventional norms or rational expectations. It is the whisper of inspiration that prompts risk-taking, opens uncharted pathways, and unveils possibilities that tools—bound by data and predefined rules—might overlook.

For example, a writer may abandon a carefully planned narrative to follow a sudden, intuitive idea, resulting in a plot twist that adds unexpected depth to their story. Similarly, an artist inspired by a fleeting impression might experiment with an unconventional color palette, creating a piece that captivates viewers with its originality. These instinctive choices often lead to breakthroughs, fostering a space where raw creativity thrives unencumbered by external constraints.

Emotion: The Soul of Creation

Emotion is the soul of creativity, transforming raw materials into living expressions of human experience. It elevates art, music, and writing from mere constructs to profound connections that transcend cultural and linguistic boundaries. Emotional authenticity gives creative works their universal appeal, enabling them to evoke feelings that resonate across time and geography.

For instance, a painter processing personal grief might create a piece that speaks directly to the vulnerability of loss, touching viewers who find their own emotions mirrored in the work. Similarly, a poet writing from a place of love or longing may craft verses that capture the shared human experience of connection and separation. Emotion imbues these works with a timeless quality, turning them into conversations that bridge the gap between creator and audience.

Creation as a Mirror of the Inner World

When creators rely on intuition and emotion, their work becomes a direct reflection of their inner world—a snapshot of their thoughts, feelings, and imagination in a specific moment. This unfiltered authenticity ensures a level of depth and originality that tools, no matter how sophisticated, cannot achieve.

For example, a musician composing a melody without the aid of instruments or technology might draw solely from the rhythm of their heartbeat or the tone of their voice. The result is music that feels deeply personal, unshaped by external influences. Similarly, a sculptor working with their hands might channel their emotions into every curve and detail, crafting a piece that is as much an expression of their identity as it is a work of art.

The Absence of Tools as a Catalyst for Authenticity

Far from hindering creativity, the absence of tools often amplifies its authenticity. Without the safety net of technological aids, creators are compelled to rely entirely on their instincts and emotions, fostering an intimacy and immediacy that tools cannot replicate.

For instance, a dancer performing without music or choreography might draw purely from their emotions, creating movements that are spontaneous and evocative. This raw, instinct-driven expression captures the essence of creativity as a lived experience rather than a calculated process. The vulnerability and honesty inherent in such performances often leave an indelible impression on audiences, reminding us of the power of unmediated human expression.

The Universality of Intuition and Emotion

While tools and techniques vary across cultures and disciplines, intuition and emotion are universal elements of creativity. They connect creators to their work and to one another, transcending boundaries of language, tradition, and medium. A painter in one part of the world and a musician in another might explore different themes, but both draw from the same well of human emotion and instinct.

This universality underscores the enduring significance of intuition and emotion in the creative process. It reminds us that creativity is not defined by perfection or efficiency but by connection—between the creator and their work, their audience, and the broader world.

Preserving the Heart of Creativity

In an age where tools and technologies increasingly dominate creative practices, intuition and emotion serve as a reminder of what makes creativity uniquely human. These qualities ensure that works remain rooted in personal experience and genuine expression, preserving the heart of creation even as methods evolve.

By embracing intuition and emotion, creators reaffirm the timeless essence of their craft, ensuring that their work resonates with authenticity and meaning. It is through these deeply human qualities that creativity continues to inspire, connect, and transform—transcending the limitations of tools and touching the soul.

Limitations as Catalysts: How Constraints Fuel Creativity

While tools often simplify and enhance creative endeavors, the absence of resources or the imposition of rigid constraints has historically catalyzed some of the most profound and innovative expressions of human ingenuity. Paradoxically, limitations do not hinder creativity; they amplify it. Constraints compel creators to adapt, innovate, and stretch the boundaries of their imagination, producing works that are resourceful, impactful, and deeply resonant.

The Creative Power of Boundaries

Boundaries provide a framework that forces creators to operate within defined parameters, sparking ingenuity and encouraging unconventional solutions. Across artistic disciplines, constraints often serve as inspiration rather than restriction, transforming the creative process into a dynamic exploration of possibilities.

For instance, visual artists working with a restricted palette must evoke emotion and depth using just a few colors. This limitation drives them to experiment with texture, composition, and form, resulting in striking works of simplicity and sophistication. Similarly, musicians constrained to a single instrument explore its full range, uncovering unique tonal qualities and crafting innovative soundscapes. These creative boundaries challenge artists to think differently, leading to discoveries that might never emerge in an unconstrained environment.

Literature and the Art of Constraint

The principle of limitation as a creative catalyst is especially evident in literature. Throughout history, poets and authors have embraced structural constraints, such as rhyme schemes, syllable counts, or fixed genres, to craft works that stand the test of time. These limitations demand precision and intentionality, encouraging creators to distill complex ideas into concise, impactful expressions.

The haiku, for example, with its strict 5-7-5 syllable structure, forces poets to convey profound imagery or emotion within just seventeen syllables. This brevity fosters clarity and depth, making the haiku a powerful medium for capturing fleeting moments or universal truths. Similarly, the sonnet—used by poets like Shakespeare—harnesses a rigid structure to explore timeless themes like love and mortality. These examples highlight how constraints, far from restricting creativity, refine and elevate it.

Early Filmmaking: Innovation Through Constraint

The early days of filmmaking offer compelling examples of how technological constraints inspire extraordinary innovation. Limited by basic cameras, static shots, and minimal editing tools, filmmakers relied on narrative strength, evocative lighting, and creative staging to captivate audiences.

Pioneers like Georges Méliès used physical props, stop-motion animation, and clever camera tricks to produce magical, otherworldly scenes that defined early cinema. Similarly, silent film actors like Charlie Chaplin conveyed emotion and humor through expressive body language, proving that creativity thrives even in the absence of sound or sophisticated equipment. These groundbreaking works demonstrate that constraints often lead to inventive techniques that redefine their medium.

The Role of Constraints in Modern Creativity

Even in today's technology-rich environment, creators frequently impose limitations on themselves to ignite innovation. Minimalist designers intentionally work with fewer elements, emphasizing clarity and impact through simplicity. In music, some artists deliberately use vintage equipment or restrict themselves to a limited number of instruments, embracing constraints to produce raw, authentic sounds.

In gaming, limitations in early hardware and software gave rise to iconic innovations. Games like *Tetris* and *Pac-Man*, developed within tight memory and processing constraints, focused on simple yet addictive gameplay mechanics. These limitations forced developers to prioritize user engagement, resulting in timeless designs that continue to influence modern gaming.

Lessons from Constraints: Embracing Resourcefulness

Working within limitations fosters resourcefulness, a cornerstone of creativity. When resources are scarce or tools are unavailable, creators are compelled to experiment, iterate, and think beyond conventional solutions. This adaptability often leads to surprising and transformative results, proving that abundance is not a prerequisite for creativity.

During wartime shortages, for instance, artists repurposed materials to create works of art and functional objects, imbuing them with unique symbolism and emotional depth. In today's sustainability-focused world, designers and engineers embrace constraints such as limited resources or environmental considerations to innovate in renewable energy, zero-waste production, and green architecture. These examples demonstrate that constraints not only challenge creators but also inspire solutions that are both meaningful and impactful.

The Gift of Limitations

While tools and technologies expand possibilities, limitations remind us of the power of the human mind to innovate and adapt. Constraints strip away excess, focusing attention on the essentials and deepening the connection between the creator and their work. By doing so, they often lead to creations that are not just innovative but profoundly resonant.

By embracing the challenges posed by constraints, creators transform obstacles into opportunities, reaffirming that the essence of creativity lies not in the tools we possess but in our ability to imagine and adapt. Constraints serve as powerful catalysts, elevating the creative process and reminding us that true innovation often emerges in the face of challenge.

This perspective underscores a universal truth: limitations, rather than hindering creativity, often unlock its fullest potential.

Why Some Creators Value Simplicity Over Sophistication

Simplicity, for many creators, represents the purest form of creativity—a deliberate choice that strips away distractions and focuses on the essence of their craft. By choosing simplicity over sophistication, these creators aim to elevate their work, deepen their connection to the creative process, and produce results that feel authentic and timeless. Far from being a limitation, simplicity becomes a catalyst for innovation, a statement of intent, and a celebration of craftsmanship.

Simplicity as a Path to Purity

For purists, sophisticated tools, with their layers of functionality and automation, risk overcomplicating the creative process. While these tools can offer convenience, they may also dilute the creator's original vision, leading to works that feel overly polished but lack soul. Simplicity, on the other hand, demands direct engagement with materials and ideas, resulting in unfiltered, genuine expressions.

This philosophy is at the core of minimalism, which celebrates simplicity in design, art, and lifestyle. Minimalist creators intentionally reduce their tools and elements, focusing instead on form, function, and meaning. By removing unnecessary details, they ensure that every component of their work serves a purpose, creating pieces that are both profound and elegant in their restraint.

The Timeless Appeal of Traditional Craftsmanship

The enduring value of simplicity is perhaps most evident in traditional craftsmanship. A potter shaping clay by hand or a carpenter crafting furniture with simple tools reflects an intimate connection to their materials and craft. These creators rely on skill, intuition, and effort to produce objects that are as beautiful as they are functional. For them, the creative process is as significant as the finished product.

In these practices, the absence of modern aids is not a shortcoming but an opportunity. The imperfections of handcrafted work—a subtle glaze variation, an irregular seam—add character and individuality, qualities often lost in mass production or automated processes. These nuances tell a story of human effort and authenticity, making each creation unique.

The Joy of Mastery Through Simplicity

Simplicity challenges creators to master their craft without reliance on advanced tools, requiring patience, practice, and a deep understanding of their medium. This journey fosters growth and discovery, as every challenge overcome enhances the creator's sense of achievement.

For example, a musician playing an acoustic guitar explores its tonal range and nuances, finding ways to evoke complex emotions with just a few chords. Similarly, a painter working with a limited palette discovers new techniques to blend and layer colors, creating richness and depth from simplicity. These constraints push creators to innovate within boundaries, resulting in works that feel both personal and profound.

Simplicity as a Statement Against Excess

In an era dominated by rapid technological advancement and consumerism, the choice to embrace simplicity can be a statement against excess. By rejecting sophisticated tools, creators reaffirm the value of intentionality, effort, and sustainability. Their work becomes a reminder that creativity does not depend on the latest technologies but on the ability to transform basic materials into something meaningful.

This perspective resonates in movements like slow design and handmade goods, which prioritize craftsmanship, quality, and sustainability over speed and scale. By advocating for thoughtful, deliberate creation, these movements challenge the notion that "more is better," emphasizing instead the importance of mindful innovation.

The Connection Between Creator and Creation

One of the most profound rewards of simplicity is the intimate connection it fosters between the creator and their work. Without advanced tools mediating the process, every decision and gesture becomes a direct expression of the creator's intent. This hands-on engagement infuses the work with authenticity and individuality, qualities that resonate deeply with both the creator and their audience.

For instance, a writer drafting with pen and paper may find that the tactile experience encourages deeper reflection and intentionality. Similarly, a sculptor carving stone by hand develops an intimate understanding of its texture and resistance, resulting in a creation that feels alive with their touch. This closeness enhances the creator's satisfaction and allows audiences to appreciate the unique effort behind the work.

The Enduring Relevance of Simplicity

In a world increasingly driven by technology, the choice to value simplicity over sophistication remains profoundly relevant. It serves as a reminder that creativity is not defined by the tools we use but by the vision, passion, and effort we bring to the process. Simplicity offers a path to authenticity, ensuring that the creator's skill and intent take center stage, unencumbered by the distractions of complexity.

By embracing simplicity, creators reconnect with the timeless essence of their craft, celebrating the human spirit and the ingenuity that arises from working with what they have. It is a philosophy that values effort over ease, meaning over embellishment, and connection over convenience. Simplicity, ultimately, reminds us that the most profound creative expressions often emerge not from what we possess, but from what we choose to create.

The Enduring Appeal of Raw Creativity

Raw creativity embodies the essence of human ingenuity—a pure expression of imagination, emotion, and resourcefulness untethered by the mediating influence of external tools. It stands as a powerful testament to the resilience and adaptability of the human spirit, illustrating that creativity often flourishes not in the absence of challenges but because of them. While tools can streamline processes or expand possibilities, purists argue that they can never replicate the authenticity, emotional depth, and personal connection that define unassisted creation.

The Power of Intuition and Emotion

At the core of raw creativity are intuition and emotion—qualities that are uniquely human and vital to the act of creation. Intuition allows creators to traverse uncharted territories, making bold leaps of imagination that logic alone might not justify. Emotion, on the other hand, infuses their work with resonance and meaning, forging a profound connection between the creator and their audience.

An artist working without advanced tools might rely entirely on instinct and feeling, channeling raw emotion into a painting or sculpture. The imperfections in such works—an uneven stroke, a rough edge—become part of their charm, reflecting the humanity of the creator and their intimate relationship with the medium. It is this emotional authenticity that gives raw creativity its enduring appeal, ensuring its relevance in an increasingly mechanized world.

The Beauty of Constraints

Constraints, often seen as limitations, are celebrated in raw creativity for their transformative potential. By imposing boundaries, constraints force creators to adapt, innovate, and maximize the resources at their disposal. This process not only sharpens their skills but also results in works that are imaginative, resourceful, and uniquely compelling.

For instance, a poet crafting profound meaning within the rigid structure of a haiku or a filmmaker creating visually stunning narratives with minimal technology demonstrates how constraints push creators to explore the full extent of their ingenuity. The absence of tools or advanced resources becomes an opportunity rather than a hindrance, fostering innovation and individuality in the creative process.

Authenticity in Simplicity

Simplicity is a hallmark of raw creativity, emphasizing the essence of the work over embellishment. By stripping away complexities, creators can focus on the heart of their craft, ensuring that their vision shines through unencumbered. This dedication to simplicity often results in creations that feel honest, grounded, and timeless.

Traditional craftsmanship serves as a profound example of this principle. A potter shaping clay by hand or a woodworker crafting furniture without power tools creates pieces imbued with character and soul. These works are not only functional but also deeply personal, reflecting the creator's effort, intention, and connection to their craft.

The Timelessness of Raw Creation

Raw creativity transcends time and technological advancement, rooted in qualities that are fundamentally human. Long before the advent of sophisticated tools, our ancestors painted on cave walls, told stories by firelight, and built awe-inspiring structures using basic means. These achievements remind us that the essence of creativity lies not in the tools we wield but in the vision, passion, and resourcefulness we bring to the process.

Even in the modern era, raw creativity holds a special place. In a world dominated by technology and automation, unassisted creation serves as a counterbalance, reminding us of the enduring value of effort, intention, and personal connection. It invites us to slow down, engage deeply with our craft, and rediscover the joy of making something with our own hands and minds.

A Human-Centered Approach to Creativity

Raw creativity celebrates the uniquely human aspects of creation—connection, emotion, and authenticity—over efficiency or perfection. It prioritizes the journey of creation, with all its challenges and discoveries, affirming that the process is as significant as the outcome.

For purists, this approach is not a rejection of tools but a reaffirmation of the essence of creativity as an intimately personal endeavor. It highlights that the most meaningful works often arise not from convenience but from effort, perseverance, and passion.

A Testament to the Human Spirit

The enduring appeal of raw creativity lies in its ability to inspire, connect, and transform. It is a testament to the resilience and adaptability of the human spirit, showing that creativity is not about resources but about the depth of the creator's inner world. By valuing simplicity, embracing constraints, and prioritizing intuition and emotion, raw creativity ensures that the essence of human ingenuity continues to shine brightly in a technology-driven world.

In choosing raw creativity, we celebrate the timeless qualities that define us as creators. It reminds us that creativity is more than a skill—it is a reflection of our humanity, a powerful force that bridges the gap between the individual and the universal, and an enduring testament to the beauty of the human spirit.

5.2 Case Studies of Creators Rejecting Tools

Throughout history and into the modern era, there have been creators who deliberately reject the use of advanced tools, favoring simplicity and raw expression over sophistication. Their work demonstrates the power of minimalism, the preservation of tradition, and the deep connection between the creator and their medium. This section explores notable examples of tool-free creativity, highlighting how these creators achieve profound outcomes through deliberate limitations.

Minimalist Art: A Celebration of Simplicity

Minimalist art epitomizes the philosophy that "less is more," emphasizing the core elements of form, color, and texture while eliminating unnecessary embellishments. Stripped of complexity, this art form highlights clarity and depth, inviting viewers to connect deeply with its essence. By rejecting elaborate techniques and tools, minimalist artists celebrate simplicity as a profound and liberating approach to creativity, where every element has significance and purpose.

Simplicity as a Statement

Minimalist art challenges conventional notions of artistic brilliance by showcasing the power of restraint. Rather than relying on intricate details or grand compositions, minimalism focuses on clean lines, muted color palettes, and open spaces that allow each element to breathe and convey meaning. This deliberate reduction strips art down to its essentials, drawing attention to the purity of the creative process and fostering a direct relationship between the viewer and the artwork.

For minimalist creators, simplicity is not a limitation but a framework for exploration. By reducing elements to their most fundamental forms, they create space for deeper reflection and engagement. This reductionist approach demands precision and intention—every decision regarding shape, texture, or placement becomes critical to the overall impact of the piece. Minimalism, in its essence, transforms simplicity into a bold statement about clarity and focus.

The Works of Agnes Martin and Donald Judd

Artists such as Agnes Martin and Donald Judd exemplify the minimalist ethos, crafting works that emphasize geometric forms and subtle variations. Martin's serene grid paintings, with their delicate lines and muted hues, are meditative explorations of balance and harmony. Her art invites quiet contemplation, guiding viewers to experience the interplay of light, space, and emotion. Each stroke reflects her disciplined yet intuitive approach, capturing the sublime in simplicity.

Donald Judd's sculptures, on the other hand, focus on materiality and spatial relationships. Crafted from industrial materials like aluminum and steel, Judd's three-dimensional works explore the purity of form and its interaction with the surrounding environment. His emphasis on simplicity elevates the ordinary, showing how even the most utilitarian shapes can hold profound meaning when approached with intention. Together, Martin and Judd demonstrate how minimalism transforms simplicity into a powerful artistic language.

The Emotional Impact of Minimalism

Despite its pared-down aesthetic, minimalist art is far from simplistic. By removing distractions, it sharpens focus on the subtle details and underlying emotions embedded within the work. Through repetition, symmetry, and delicate variations, minimalist pieces create rhythms that draw viewers into deeper engagement.

For example, a minimalist painting consisting of a single block of color or a few carefully placed lines might evoke feelings of calm, introspection, or tension. Its apparent simplicity encourages viewers to explore nuances in texture, balance, and spatial relationships. Without visual clutter, minimalist art invites intellectual and emotional reflection, offering a contemplative and impactful experience.

The Connection Between Artist and Viewer

Minimalist art fosters a unique and intimate connection between the artist and the viewer. By focusing on essentials and embracing simplicity, minimalist works leave ample space for personal interpretation. This open-ended quality invites viewers to bring their own perspectives, emotions, and experiences to the piece, transforming the artwork into a shared dialogue.

For the artist, this connection is central to the minimalist philosophy. The absence of elaborate techniques ensures that the artwork's core message remains unencumbered and accessible. This purity of intent allows the viewer to engage with the piece on their own terms, finding meaning in its clarity and simplicity. Minimalism thus becomes a bridge between creator and audience, rooted in shared reflection and understanding.

Minimalism as a Timeless Philosophy

Minimalist art endures as a timeless and influential movement, demonstrating that simplicity can be as powerful and evocative as complexity. Its enduring appeal lies in its ability to distill the essence of human experience into forms that are both profound and accessible. By rejecting excess and embracing restraint, minimalist artists create works that transcend fleeting trends, standing as enduring symbols of clarity and focus.

In a world increasingly dominated by noise and excess, minimalism offers a refreshing counterpoint—a perspective that values intention, clarity, and the quiet power of less. It reminds us that the most impactful creations often arise not from abundance, but from thoughtful reduction. By celebrating simplicity, minimalist art affirms that beauty lies not in how much we add, but in how much we allow ourselves to see.

Traditional Craftsmanship: Preserving Handcrafted Excellence

Traditional craftsmanship represents the timeless value of creating with minimal tools or entirely by hand, celebrating artistry, skill, and dedication over the conveniences of modern machinery. This approach honors the human element in creation, producing works that are deeply personal and enriched with cultural and historical significance. By relying on expertise and an intimate connection with materials, traditional craftsmen preserve the authenticity of their craft, ensuring that every piece tells a unique and meaningful story.

The Essence of Handcrafted Mastery

At the heart of traditional craftsmanship is a reliance on human effort, intuition, and expertise. Craftsmen devote years to mastering their techniques, refining their skills through practice and dedication. This mastery elevates everyday objects into works of art, where imperfections are not flaws but markers of individuality. Each piece carries the artisan's fingerprint, a tangible connection to the creator's vision and skill.

These artisans approach their materials with reverence, understanding the distinct characteristics of wood grain, clay pliability, or the texture of woven fibers. This deep knowledge ensures that the final product reflects both the natural beauty of the materials and the creative intent of the maker. The process itself becomes a dialogue between the creator and their medium, producing objects that resonate with authenticity and timeless charm.

Japanese Joinery: A Testament to Precision and Tradition

Japanese woodworkers exemplify traditional craftsmanship through their intricate and celebrated art of joinery. This technique involves interlocking wooden pieces without nails, screws, or adhesives, showcasing an extraordinary blend of precision, creativity, and structural integrity. Each joint is meticulously crafted to provide strength while enhancing the wood's natural beauty, reflecting a philosophy where function and form harmonize seamlessly.

Rooted deeply in Japanese culture, joinery techniques have been passed down for generations. Master craftsmen possess an intimate understanding of wood properties, such as its response to changes in humidity or temperature, allowing them to create enduring pieces. From Zen-inspired temples to finely crafted chests, Japanese joinery symbolizes the enduring power of traditional techniques in a world often driven by mass production.

Pottery: The Touch of the Hand

In pottery, the human touch is central to the creative process, with each hand-thrown ceramic piece bearing the potter's unique imprint. From shaping the clay on a wheel to firing it in a kiln, the process relies as much on intuition as on technical skill. No two pieces are ever identical, with natural variations and imperfections adding character and charm.

Traditional potters embrace these variations, viewing them as hallmarks of authenticity. Unique glaze patterns, slight asymmetries, or subtle shifts in form reflect the organic nature of handcrafted ceramics, distinguishing them from their machine-made counterparts. Beyond their aesthetic appeal, these works preserve centuries-old techniques and cultural traditions, ensuring their relevance in both function and artistry.

Weaving and Textiles: Threads of Tradition

Traditional weaving practices highlight the enduring value of handcrafted textiles, with artisans around the world preserving intricate methods passed down through generations. In regions such as India, Africa, and South America, handloom weavers create fabrics imbued with cultural significance, often incorporating patterns and techniques unique to their communities.

For instance, Peruvian weavers use natural dyes and hand-spun fibers to craft vibrant, symbolic patterns that reflect their connection to the Andes and their ancestors. These textiles serve as both functional items and cultural artifacts, embodying stories of heritage, identity, and artistry. The painstaking attention to detail and respect for tradition ensure that each piece is a testament to the weaver's dedication and skill.

The Significance of Preserving Tradition

Traditional craftsmanship represents more than a method of creation—it is a philosophy that values authenticity, patience, and cultural continuity. By resisting the convenience of modern tools, these artisans maintain a profound connection to their craft, ensuring that their work is deeply rooted in heritage. Their creations serve as a reminder of the importance of slowing down, valuing quality over quantity, and appreciating the stories behind the objects we use and treasure.

In an era dominated by industrialization and digital innovation, traditional craftsmanship offers a counterpoint to mass production. It emphasizes the human connection to creation, celebrating the individuality and meaning infused in every handmade object. These works are not just products—they are reflections of creativity, history, and shared humanity.

A Legacy of Handcrafted Excellence

The enduring appeal of traditional craftsmanship lies in its ability to preserve the human touch in an increasingly mechanized world. By honoring the past and inspiring future generations, these artisans ensure that their creations remain timeless symbols of human ingenuity and cultural heritage. Embracing simplicity, patience, and dedication, traditional craftsmen reaffirm the value of authenticity and the beauty of handcrafting, keeping alive a legacy that continues to enrich and inspire.

Writers and Thinkers Who Avoid Technology

In a digital age where advanced tools dominate the creative process, a distinct group of writers and thinkers consciously distance themselves from modern technology. Choosing simplicity over sophistication, these creators rely on traditional methods such as handwriting, typewriters, or basic word processors. Their decision is rooted in the belief that creativity flourishes best when the process remains organic and unencumbered, fostering a deep connection to their ideas and ensuring that their work retains its authenticity.

The Intimacy of Handwriting

For many writers, handwriting represents a unique and deeply personal way to engage with their work. The tactile act of putting pen to paper slows the creative process, allowing ideas to flow naturally while encouraging reflection. This deliberate pace fosters mindfulness, free from the distractions of formatting tools, pop-up notifications, or the ever-tempting internet.

Celebrated authors like J.K. Rowling have emphasized their preference for handwriting drafts, describing how it enables them to immerse themselves in their characters and storylines. Rowling's handwritten pages, often brimming with annotations and sketches, exemplify the intimate connection between writer and text. For these creators, the physicality of handwriting transforms the blank page into a canvas for unfiltered thought and raw emotion, making the process itself as meaningful as the final product.

The Nostalgia and Focus of Typewriters

Typewriters, though relics of a bygone era, remain beloved tools for many writers who value their simplicity and tactile charm. Free from the distractions of digital technology, typewriters offer a focused and immersive writing experience. The rhythmic sound of keystrokes and the permanence of ink on paper create a sense of deliberation and immediacy, grounding the writer in the present moment.

Icons like Jack Kerouac and Hunter S. Thompson famously relied on typewriters to produce their groundbreaking works. The inability to delete or endlessly revise encouraged a raw and spontaneous style, capturing thoughts as they emerged. The physical act of typing on a machine, where each decision feels final, lends a sense of authenticity to the writing process, demonstrating a commitment to storytelling unfiltered by technological convenience.

The Minimalism of Basic Word Processors

In an effort to balance tradition with technology, some contemporary writers opt for basic word processors that strip away the distractions of modern software. George R.R. Martin, creator of *A Song of Ice and Fire*, exemplifies this approach. Martin writes his epic novels on a DOS-based word processor devoid of internet access, spellcheck, or formatting tools. This minimalist interface enables him to focus entirely on his fictional world, undisturbed by external interruptions.

For Martin and others like him, simplicity is not a limitation but a tool to enhance immersion. By eliminating unnecessary features, they create an environment conducive to deep focus, ensuring that the essence of their storytelling remains intact. This intentional choice underscores their dedication to the craft and highlights the enduring appeal of low-tech solutions in a high-tech world.

The Benefits of Rejecting Technology

Choosing to avoid advanced tools is not an act of nostalgia or defiance but a deliberate strategy to nurture focus and creativity. Modern technology, while powerful, often introduces distractions that fragment attention and disrupt the flow of ideas. Writers and thinkers who embrace traditional methods create a space where clarity and depth take precedence, allowing them to connect more meaningfully with their work.

The slower pace of handwriting or typewriting, for instance, encourages thoughtful decision-making and a stronger engagement with each word. This mindfulness enriches the creative process, resulting in stories and ideas that feel authentic and considered. For these creators, the journey of crafting their work becomes an act of reflection and discovery, yielding results that resonate on a deeper level.

Preserving the Organic Process of Storytelling

At its core, the decision to reject advanced tools reflects a commitment to the organic process of storytelling. Writers who embrace traditional methods prioritize the raw and unpolished nature of their early drafts, valuing authenticity over efficiency. By working within the limitations of simpler tools, they maintain an unmediated connection to their ideas, ensuring that their creative vision remains pure and true.

This philosophy serves as a counterpoint to the rapid pace of contemporary creation, emphasizing that meaningful work is not about speed or convenience but about depth and intentionality. Whether through handwritten notes, typewritten pages, or basic digital interfaces, these creators reaffirm the value of immersive and thoughtful engagement with their craft.

A Testament to Intentional Creativity

The choice to work without advanced technology is not about rejecting progress but about preserving the essence of intentional creativity. By focusing on the rhythm and flow of their thoughts, these writers demonstrate that profound work often arises from simplicity and deliberate effort. Their approach challenges the tech-driven norms of modern life, offering an alternative perspective that prioritizes connection over convenience.

In an era where technology dominates the creative landscape, these writers stand as champions of raw, human-centered creation. Their work is a testament to the enduring power of simplicity, proving that the most impactful results often emerge not from what we have but from how we choose to use it. By embracing methods that prioritize authenticity, these creators remind us that the heart of storytelling lies in its unfiltered connection to the human experience.

Performers and Artists Embracing Raw Expression

In the realms of performance art and music, some creators purposefully reject advanced tools and technology, opting instead for unfiltered, raw expression. This approach prioritizes the emotional connection between performer and audience, emphasizing spontaneity, authenticity, and the immediacy of live interaction. By stripping away external enhancements, these artists allow their work to speak for itself, fostering a profound sense of intimacy and resonance that is rare in highly produced settings.

The Power of Raw Musical Expression

For many musicians, simplicity becomes a deliberate choice, one that amplifies the emotional depth and authenticity of their performances. Folk musicians, for instance, rely on acoustic instruments or the pure resonance of their voices, rejecting the need for amplification or digital effects. This focus on natural sound highlights subtle nuances and timbres, drawing listeners into the heart of the music.

Artists in genres such as blues, folk, and classical music often embody this raw approach. A singer performing a capella depends entirely on their voice to convey emotion and meaning, establishing a deeply personal connection with the audience. Similarly, an acoustic guitarist's melodies, played without amplification or effects, evoke a sense of intimacy and nostalgia. This unadorned artistry resonates deeply with audiences, who feel they are experiencing something genuine, personal, and unrepeatable.

Dance as an Expression of the Unfiltered Self

In the world of contemporary dance, raw expression often takes center stage. Pioneering choreographers such as Pina Bausch rejected elaborate stage designs and technical embellishments, instead focusing on the body's natural movements to convey complex emotions. Her performances emphasized vulnerability and the raw intensity of human experience, allowing audiences to witness unfiltered physical storytelling.

This approach extends to improvisational and experimental dance forms, where the absence of digital projections, pre-programmed lighting, or other technological aids shifts the emphasis entirely to the dancer's movement. Each step, pause, and gesture becomes a profound narrative element, creating a visceral connection between the performer and their audience. Without external distractions, the emotional power of the movement itself becomes the focal point, inviting viewers into a shared, immediate experience.

The Role of Spontaneity in Live Performances

Live performances that forgo advanced tools often thrive on spontaneity, transforming each show into a unique and unpredictable experience. By rejecting technological intervention, performers are free to respond in real time to their environment, emotions, and audience. This dynamic interplay creates an authenticity and energy that cannot be replicated in highly produced settings.

Street performers and improvisational theater artists, for example, thrive on the immediacy of their surroundings. They adapt their acts to the mood of the crowd, the acoustics of the space, and even unexpected disruptions. Every performance becomes a one-of-a-kind experience, with each moment—planned or spontaneous—woven into the fabric of their art. This raw, unscripted quality ensures that the performance feels alive and memorable, for both the artist and the audience.

The Intimacy of Tool-Free Expression

The deliberate rejection of technological aids often results in an unparalleled intimacy between the performer and the audience. When artists rely solely on their own abilities, viewers become more attuned to the subtleties of their craft—the delicate inflections of a singer's voice, the expressive pauses in a dancer's movements, or the raw emotion in an actor's breath. These nuances create a deep connection that transcends the physical space between the artist and their audience.

This intimacy is especially evident in unplugged sessions, house concerts, and community theater performances. In these smaller, unamplified settings, the absence of technological mediation allows the art to unfold naturally.

The shared experience becomes one of trust and presence, where both performer and audience engage in a dialogue that feels genuine and immediate.

The Philosophical Choice of Raw Expression

For many performers and artists, embracing raw expression is as much a philosophical stance as it is an artistic decision. By rejecting technological aids, they underscore the importance of human effort, authenticity, and vulnerability in the creative process. This perspective challenges the prevailing notion that perfection is the ultimate goal of art, instead celebrating the imperfections and unpredictability that make each performance distinct.

In an era dominated by digital enhancements and polished productions, this choice stands as a powerful counterpoint. It reminds us that art is not solely about the final product but about how it is created and experienced. The raw, unfiltered nature of these performances asserts that imperfection is not a flaw but a hallmark of genuine creativity.

A Celebration of the Human Connection

Artists and performers who embrace raw expression reaffirm the timeless power of human creativity. By focusing on unmediated, authentic interaction, they transcend the limitations of tools and technology, proving that art's true essence lies in its ability to connect, inspire, and move. Their work, unadorned yet deeply impactful, celebrates the beauty of imperfection and the enduring appeal of simplicity.

Through their choice to prioritize authenticity over convenience, these creators ensure that their art remains deeply human. They remind us that, at its core, creativity is about forging connections—between the artist and the audience, the performer and the moment, and ultimately, the human spirit and its boundless capacity for expression.

Lessons from Tool-Free Creativity: Balancing Innovation and Tradition

Creators who deliberately avoid advanced tools remind us of the enduring value of simplicity, intuition, and human effort in the creative process. Their work demonstrates that while tools can amplify our capabilities, they are not the essence of creativity. Instead, true creativity lies in the ability to connect deeply with the medium, the message, and the audience. By balancing innovation with tradition, these creators show us how to navigate a world increasingly dominated by technology without losing touch with our human roots.

Grounding Innovation in Human Effort

Tool-free creativity underscores that innovation originates not from tools but from the vision, skill, and perseverance of the creator. While advanced technologies can enhance the creative process, they are ultimately shaped by the human ingenuity behind them. By rejecting modern tools, some creators illustrate the power of relying on intuition and effort, producing deeply personal and resonant works.

For example, a sculptor carving stone by hand relies on technical skill and an intimate connection with the material. This hands-on approach fosters a sense of accomplishment and authenticity that is difficult to replicate with mechanized tools. Similarly, writers drafting by hand or on typewriters often describe the process as more focused and immersive, allowing them to engage deeply with their ideas.

The Relevance of Simplicity in a Complex World

In a fast-paced, technologically advanced world, the choice to embrace simplicity serves as a counterpoint to complexity. Tool-free creators highlight that impactful innovation does not always require sophisticated resources. Often, the most meaningful creations emerge from slowing down, stripping away distractions, and focusing on the core elements of the craft.

This approach is particularly evident in fields like art and music, where the human touch is integral to emotional resonance. Minimalist artists, for instance, distill their ideas into pure, unembellished forms, creating works that deeply connect with audiences. This philosophy challenges the assumption that more technology equals better results, showcasing the enduring power of thoughtful, deliberate creation.

Balancing Tradition with Modernity

Tool-free creators often draw inspiration from traditional methods, preserving techniques refined over generations. This commitment to tradition bridges the past and present, ensuring that the wisdom of earlier eras informs contemporary practices. At the same time, these creators integrate modern ideas without compromising authenticity, striking a balance between heritage and innovation.

For instance, Japanese woodworkers practicing the art of joinery create intricate designs without nails or screws, relying solely on interlocking pieces. While rooted in tradition, their work incorporates modern aesthetics, demonstrating that innovation and tradition can coexist. This harmonious blend ensures that their craft remains relevant while honoring its origins.

Authenticity in Connection with the Audience

One of the most compelling lessons from tool-free creativity is its emphasis on authenticity as a bridge to the audience. Creations born without advanced tools often carry a rawness and vulnerability that resonate deeply. This authenticity stems from the creator's direct engagement with their medium, free from technological mediation.

For instance, acoustic musicians performing live captivate audiences with the unadorned purity of their sound. Similarly, handmade crafts evoke nostalgia and appreciation for the effort involved, strengthening the bond between creator and consumer. This raw, unfiltered quality enhances the emotional impact of the work, creating a meaningful connection.

A Counterbalance to Technological Overdependence

In an era of pervasive technology, tool-free creators offer a counterbalance, reminding us of the risks of overreliance on tools. While technology can enhance creativity, it can also distance us from the process of making. By choosing simplicity, creators reclaim the joy of hands-on creation, emphasizing that the journey of making is as important as the outcome.

This perspective fosters a mindful approach to creativity, prioritizing authenticity over efficiency. It challenges the notion that speed and output are the ultimate goals, celebrating instead the discovery, growth, and fulfillment inherent in the creative process.

Honoring Tradition and Cultural Heritage

Tool-free creativity also serves to preserve cultural traditions and heritage. By continuing age-old practices, creators ensure that the knowledge and skills of their ancestors remain relevant. Traditional textile weavers, potters, and woodworkers, for example, reflect the values and aesthetics of their communities, acting as custodians of cultural identity.

These practices highlight the importance of storytelling through craft, offering a connection to the past that enriches the present. By honoring tradition, these creators provide a framework for future generations to balance heritage with innovation.

Inspiration for Modern Creators

The legacy of tool-free creativity challenges modern creators to reflect on their relationship with technology. It poses essential questions: Are we overly dependent on tools? Are we sacrificing authenticity for convenience? How can we maintain integrity while embracing innovation?

By valuing simplicity, tradition, and human effort, creators can balance technological advances with the authenticity of their craft. This mindful approach ensures that their work remains meaningful and grounded, inspiring a deeper connection to the creative process.

A Timeless Lesson for Creativity

The rejection of tools is not a rejection of progress but a celebration of timeless qualities that define human creativity: imagination, adaptability, and the desire for authentic expression. Tool-free creators remind us that the act of creation is as significant as the final product and that constraints can inspire extraordinary outcomes.

Their work challenges the modern obsession with perfection and efficiency, emphasizing the value of patience, effort, and intentionality. It calls us to reconnect with the foundational aspects of creativity, ensuring that our work remains both impactful and deeply human.

The Beauty of Unfiltered Expression

At its core, tool-free creativity celebrates the unfiltered expression of human ingenuity. It highlights the beauty of imperfection, the power of intuition, and the transformative potential of simplicity. These creators demonstrate that impactful work does not require advanced resources—only a deep connection to the process and a commitment to authenticity.

As we navigate a technology-driven world, the lessons of tool-free creativity remind us of the essence of what it means to create. By embracing simplicity and honoring tradition, we can ensure that our innovations remain true to their purpose, preserving the authenticity and depth that define meaningful art and expression.

Summary

This chapter explores the philosophy of creativity as an unfiltered, deeply human process, untainted by external tools or technologies. It presents the purist's viewpoint that true creativity arises from intrinsic capabilities like imagination, emotion, and experience, emphasizing the importance of preserving these qualities in the machine age.

Key themes include:

The Essence of Pure Creativity: Creativity is framed as an expression of the human spirit, independent of external aids. The chapter underscores the importance of raw skills, intuition, and emotional depth as the core of authentic creative processes.

Challenges of Tool Dependency: It critiques the over-reliance on tools, arguing that they can dilute originality and create a sense of creative complacency. Examples include traditional art forms like painting or poetry, where minimal external intervention highlights the creator's innate abilities.

The Role of Constraints in Creativity: Constraints are celebrated as catalysts for innovation, forcing creators to think deeply and push boundaries without external shortcuts. Historical examples illustrate how limitations have spurred breakthroughs in art, literature, and science.

Cultural and Philosophical Foundations: The chapter ties the purist approach to various cultural and philosophical traditions that value simplicity, introspection, and self-reliance. It reflects on the enduring relevance of these traditions in the face of modern technological advancements.

Modern Implications and the Balance: While advocating for purist ideals, the chapter acknowledges the challenges of adhering strictly to them in a tech-driven world. It invites readers to consider striking a balance, integrating tools where necessary while retaining the authenticity of their creative intent.

This chapter serves as a thought-provoking counterpoint to technology-driven perspectives, reminding readers of the irreplaceable value of unmediated human creativity. It challenges creators to reflect on their practices and find ways to honor the purist ethos in their work.

6: The Dilemma of Modern Creators

In an era where creativity is both amplified and challenged by rapid technological advancements, modern creators find themselves navigating a complex landscape. The tools at their disposal are more powerful than ever, enabling unprecedented innovation and efficiency. Yet, these advancements also pose unique dilemmas, forcing creators to balance the authenticity of traditional methods with the opportunities provided by cutting-edge technologies. Beyond external tools, creators must confront internal struggles—self-doubt, perfectionism, and societal expectations—that can stifle their artistic freedom. This chapter delves into the paradoxical challenges faced by today's creators, exploring how they can reconcile tradition with technology while breaking free from the restrictions that hinder their potential.

6.1 Balancing Tradition and Technology

Modern creators stand at a crossroads, navigating the intricate interplay between time-honored traditions and the allure of groundbreaking technology. This balancing act defines much of today's creative landscape, as artists, writers, musicians, and innovators strive to honor their heritage while embracing the transformative potential of modern tools. This section explores the opportunities and challenges of merging the old with the new, highlighting the strategies and stories of those who successfully harmonize tradition and technology.

The Intersection of Heritage and Innovation

Heritage and innovation, while seemingly opposites, often function as complementary forces in the creative process. Heritage offers a foundation rich in tradition, technique, and cultural narratives that ground creative work in authenticity. Innovation, by contrast, propels creators to adapt and evolve, responding to the shifting needs and possibilities of the present. Successfully balancing these two forces demands both respect for the past and a vision for the future, resulting in work that is both rooted and forward-looking.

Heritage: The Anchor of Creativity

Heritage serves as a repository of accumulated wisdom, offering creators timeless methods and stories that have shaped cultural identities over generations. Traditional practices—whether in art, music, architecture, or craftsmanship—carry with them a sense of continuity and belonging. They provide a framework upon which modern creators can build, ensuring that their work is imbued with depth and meaning.

For instance, classical music traditions like Indian ragas or Western symphonies reflect centuries of refinement, forming a rich canvas for reinterpretation. Similarly, Japanese pottery techniques such as raku emphasize imperfection and spontaneity, inspiring modern ceramicists to embrace authenticity in their craft. These traditions not only honor the past but also remind creators of the enduring value of simplicity, intentionality, and connection to roots.

Innovation: The Drive Toward the Future

While heritage anchors creativity, innovation propels it forward. Experimentation, adaptation, and reimagining allow creators to respond to changing contexts, technological advancements, and evolving audience expectations. Innovation ensures that heritage remains dynamic, evolving rather than becoming stagnant.

Filmmakers, for example, use digital tools to recreate historical settings with remarkable accuracy, blending traditional storytelling techniques with modern visual effects. Similarly, fashion designers merge traditional textiles and motifs with futuristic silhouettes, creating garments that celebrate cultural history while appealing to contemporary sensibilities. Innovation breathes new life into heritage, ensuring it remains relevant and resonant in an ever-changing world.

The Synergy of Heritage and Innovation

The true power of creativity emerges at the intersection of heritage and innovation. This synergy allows creators to draw inspiration from the past while pushing boundaries, producing work that feels both rooted in tradition and revolutionary in its execution.

Consider contemporary musicians who blend folk instruments with electronic production, weaving ancient melodies with modern beats to create compositions that are both timeless and cutting-edge. Similarly, architects who integrate ancient building techniques into sustainable designs demonstrate this balance, using traditional methods to meet modern environmental challenges. This harmonious interplay ensures that creations resonate with diverse audiences, transcending cultural and temporal divides.

Preserving Cultural Continuity

Balancing heritage and innovation is essential to preserving cultural continuity in a globalized world. Creators who incorporate traditional elements into modern works maintain the relevance of cultural practices while allowing them to evolve. This approach safeguards traditions from obsolescence, enabling them to contribute to a shared sense of identity and belonging.

For instance, indigenous artists who use digital platforms to showcase traditional crafts bring their heritage to a global audience while adapting to modern demands. Similarly, chefs who reinterpret ancestral recipes with contemporary techniques celebrate culinary heritage while catering to modern tastes. These practices demonstrate how creators can honor their roots while embracing the opportunities of the present.

A Bridge Across Generations

The intersection of heritage and innovation acts as a bridge, connecting the wisdom of the past with the aspirations of the future. Creations that successfully blend these elements resonate across generations, fostering mutual respect and understanding between traditionalists and modernists.

In literature, for example, authors who weave ancient myths into contemporary narratives create stories that feel timeless yet relevant. Visual artists who reinterpret traditional motifs using modern mediums achieve a similar effect, offering works that resonate with both heritage enthusiasts and contemporary audiences. These efforts foster a dialogue that bridges time, creating a shared space for appreciation and learning.

Crafting Timeless Creativity

Balancing heritage and innovation presents challenges, as creators risk either diluting tradition or stifling progress. However, when approached thoughtfully, this intersection becomes a wellspring of creativity, offering limitless possibilities for meaningful work.

By honoring the past while embracing the future, creators craft works that are relevant, authentic, and transformative. This dynamic interplay ensures that creativity remains a bridge across time and culture, fostering connections and inspiring generations to come.

A Dynamic Meeting Point

The intersection of heritage and innovation is not merely a balance but a dynamic space where creativity flourishes. It is here that tradition informs modernity and innovation revitalizes heritage, creating works that carry the wisdom of the past while addressing the needs of the present.

In this space, creators reaffirm the timeless essence of their craft, proving that the most impactful creations are those that honor their roots while fearlessly embracing the possibilities of the future. This interplay reminds us that creativity, at its core, is about connection—between past and present, creator and audience, tradition and progress.

Integrating Traditional Techniques with Modern Tools

The fusion of traditional techniques with modern tools offers a harmonious blend of heritage and innovation, unlocking new dimensions of creative expression. This approach seeks not to replace time-honored practices but to enhance them, allowing creators to preserve the authenticity of their craft while leveraging the efficiency and precision of contemporary technology. By combining the tactile, human-centric essence of tradition with the innovative capabilities of modern tools, creators expand their possibilities while staying rooted in their cultural and artistic heritage.

Preserving Tradition While Embracing Modernity

Integrating modern tools into traditional practices enables creators to preserve the core values and authenticity of their craft while adapting to contemporary demands. Digital tools like 3D printers, laser cutters, and design software provide opportunities to streamline processes and explore new ideas without compromising the personal touch that defines traditional work.

For example, potters can use computer-aided design (CAD) software to draft intricate patterns and forms, which are then translated into clay through skilled handcrafting. This combination retains the unique imperfections and warmth of handmade pottery while incorporating the precision of digital planning. Similarly, calligraphers might use digital templates to experiment with layouts before applying ink and brush, blending precision with the traditional artistry of hand-drawn scripts.

Expanding Creative Possibilities

Modern tools open doors to creative possibilities that may have been unattainable with traditional methods alone. This integration encourages experimentation, allowing creators to push boundaries while maintaining the principles of their craft.

Textile artisans, for instance, can use digital platforms to redesign traditional motifs, exploring innovative color combinations or patterns. Once finalized, these designs can be handwoven to preserve the intricate textures and cultural significance of the piece. Sculptors, too, benefit from using 3D modeling software to test proportions and structures before turning to traditional carving techniques in stone or wood.

By bridging manual craftsmanship and digital innovation, creators achieve a synergy that honors both the depth of tradition and the expansiveness of modern possibilities. This blend enhances their work, ensuring cultural authenticity while fostering fresh perspectives.

Efficiency Without Compromising Artistry

One of the most significant advantages of modern tools is their ability to enhance efficiency, allowing creators to focus on the most meaningful aspects of their craft. Tasks that once demanded immense time and effort—such as drafting designs, calculating dimensions, or sourcing materials—can now be streamlined through technology, freeing up time for creators to refine their artistry.

For instance, muralists can use digital tools to create detailed mockups that ensure accuracy in proportion and scale, reducing the likelihood of errors. This preparation allows them to concentrate on the expressive qualities of their work during execution. Similarly, jewelry designers can use laser engraving to incorporate intricate patterns, saving time while enhancing craftsmanship.

This balance between efficiency and artistry demonstrates that technology can complement rather than overshadow the creative process, ensuring authenticity while embracing relevance.

Retaining the Human Touch

A critical challenge in integrating modern tools with traditional practices is ensuring that the human touch—the emotional depth and individuality central to traditional craftsmanship—remains at the forefront. Technology should amplify, not replace, the connection between creator and creation.

For example, a painter might sketch concepts on a digital tablet but rely on traditional brushes and paints for the final piece, preserving the texture, spontaneity, and depth that define hand-painted art. Similarly, a musician might use composition software for arranging melodies but perform live using traditional instruments, retaining the intimacy and nuance of their craft.

This thoughtful integration ensures that technology enhances the creator's vision without diminishing the emotional authenticity that resonates with audiences.

A Dynamic Blueprint for the Future

The integration of traditional techniques with modern tools provides a blueprint for the future of creativity, demonstrating that heritage and innovation can coexist in enriching and transformative ways. This approach honors the past while addressing the opportunities and demands of the present, creating works that are both timeless and forward-thinking.

In fields ranging from fashion to architecture, this synergy is already reshaping creative processes. Fashion designers combine traditional textiles with futuristic designs, while architects incorporate ancient building techniques into sustainable modern structures. These examples highlight how the fusion of tradition and innovation produces meaningful, impactful results.

By embracing this balance, creators reaffirm that the most enduring and profound creations arise from the intersection of diverse perspectives, where heritage informs innovation and technology amplifies human ingenuity.

Honoring the Past, Inspiring the Future

The fusion of traditional techniques with modern tools celebrates the resilience and adaptability of creativity. It reflects a commitment to preserving cultural heritage while exploring new horizons, ensuring that creativity remains both authentic and dynamic. By integrating the best of the past with the possibilities of the present, creators not only elevate their craft but also contribute to a legacy that bridges time, culture, and technology.

This thoughtful interplay between tradition and innovation serves as a reminder that progress does not require abandoning the past. Instead, it invites us to honor our roots while imagining and creating a future where the essence of human creativity continues to thrive.

Preserving Authenticity in a Digital Age

In today's digital-first world, where content is created, consumed, and replicated at an unprecedented pace, preserving authenticity has become a critical challenge for creators. While technology offers remarkable convenience, efficiency, and access to broader audiences, it also risks diluting the personal touch, emotional depth, and individuality that define authentic art. Striking a balance between leveraging technological advancements and staying true to one's vision is now essential for creators across all disciplines.

The Challenge of Digital Saturation

The democratization of creativity through digital platforms has enabled countless individuals to share their work with the world. However, this flood of content often leads to homogenization. Algorithms prioritize trends, popularity, and rapid production over originality, pressuring creators to conform to prevailing styles to remain visible.

This environment can stifle innovation, with creators tempted to prioritize quantity over quality. For example, social media rewards frequent posting, encouraging creators to produce rapidly rather than meaningfully. However, creators who resist these pressures and remain committed to their unique vision often produce work that stands out for its depth and originality. Their work resonates because it prioritizes authenticity over algorithmic appeal, offering a refreshing alternative in a crowded digital landscape.

Technology as a Tool, Not a Substitute

A key principle for preserving authenticity in the digital age is recognizing technology as a tool that enhances creativity, not a substitute for it. Advanced tools like generative AI, digital design software, and editing platforms can inspire new ideas and streamline workflows, but their role should remain supportive, not dominant.

For instance, digital tools can simplify complex processes—such as generating intricate patterns for digital art or simulating effects in seconds that would take hours manually. However, the final piece gains authenticity when infused with the creator's personal touch: deliberate imperfections, nuanced color choices, or unique compositions. Creators who thoughtfully apply technology ensure that it amplifies their vision rather than replacing it, striking a balance between efficiency and individuality.

The Role of Human Emotion and Intuition

Authenticity stems from human elements like emotion, intuition, and imperfection—qualities that technology cannot replicate. These elements form the core of creativity, creating a connection between creators and their audiences that transcends technical perfection.

For example, a songwriter might use AI to generate chord progressions but pour personal experiences and emotional truths into the lyrics and melody. Similarly, a photographer might use editing software to enhance an image but relies on their intuition to capture a fleeting, meaningful moment. In such cases, the human dimension drives the work's impact, with technology serving as a supportive layer rather than the foundation.

Authenticity in Generative and AI Art

The rise of generative art and AI-driven creativity has brought new opportunities and challenges. These technologies can produce visually stunning, emotionally evocative, or technically precise works, but their ease of production raises questions about originality and authenticity.

Creators working with AI often emphasize collaboration, ensuring their vision remains central to the process. For example, a digital artist might use generative algorithms to create a base design, refining it manually to reflect their unique style and emotional intent. Similarly, writers may use AI to brainstorm ideas or generate outlines but shape the final narrative with their voice and perspective. This collaborative approach integrates innovation without compromising authenticity, ensuring the work feels both cutting-edge and deeply personal.

Demonstrating That Technology and Authenticity Can Coexist

Creators who skillfully integrate technology into their work prove that authenticity and innovation can coexist. By maintaining a clear sense of purpose, they use tools to expand their creativity while keeping their unique voice intact. This approach is particularly evident in hybrid projects that combine traditional and digital methods, producing results that are both innovative and authentic.

For example, a painter might begin their process with digital sketches but complete the piece using physical brushstrokes, blending the precision of technology with the tactile intimacy of traditional art. Similarly, a musician might use AI to experiment with harmonies but rely on live performance to convey the raw emotion of their work. These creators demonstrate that technology, when used thoughtfully, can enhance rather than detract from authenticity.

A Thoughtful Approach to Technology

Preserving authenticity in the digital age requires a deliberate, mindful approach to technology. Creators must continuously reflect on how and why they use tools, ensuring alignment with their values and goals. Embracing imperfection, individuality, and emotional resonance as central to their work allows creators to maintain their authenticity in a tech-driven world.

This perspective reframes technology as a neutral force, capable of either enhancing or diluting creativity depending on its application. Creators who stay true to their vision and prioritize their unique qualities ensure their contributions remain meaningful, resonant, and enduring.

Authenticity as a Timeless Quality

In a fast-paced digital world, authenticity remains a timeless and essential quality. It is the vulnerability, passion, and individuality of human creativity that gives work its lasting impact. By preserving this essence while navigating the challenges and opportunities of the digital age, creators continue to produce work that inspires and connects on a profound level.

Authenticity is not only possible in a technology-driven era but vital. When creators prioritize their unique vision and embrace the human elements of their craft, they demonstrate that technology and authenticity can thrive together—ensuring that the art of creation remains deeply personal, meaningful, and transformative.

The Risks and Rewards of Technological Adoption

The integration of new technologies into creative processes offers a profound duality: a gateway to boundless potential and a set of challenges that demand careful navigation. On one hand, advanced tools enable creators to achieve unprecedented precision, efficiency, and reach. On the other, they risk fostering over-reliance, ethical dilemmas, and the dilution of originality. Creators must evaluate the opportunities and challenges of technological adoption thoughtfully, ensuring their work remains innovative, authentic, and meaningful.

Unlocking Creative Potential

Modern technologies have transformed the creative landscape, empowering creators to push boundaries and reimagine what's possible. Tools like AI, generative design software, and high-performance computing provide creators with the means to explore new ideas, iterate rapidly, and experiment with groundbreaking concepts.

For example, architects can use generative design software to produce and test multiple structural iterations, refining their designs for functionality and aesthetics in real time. Similarly, musicians can leverage AI-generated harmonies and rhythms to craft entirely new genres of music, blending traditional techniques with futuristic sounds. These advancements amplify human ingenuity, enabling creators to transcend conventional limits and focus on deeper conceptual and artistic pursuits.

Streamlining Workflows

Technological tools have also revolutionized workflows, streamlining time-intensive tasks and allowing creators to dedicate more energy to the core aspects of their craft. Automation of repetitive processes—such as photo editing, video rendering, or data analysis—frees creators to focus on innovation and detail.

For instance, filmmakers can now use advanced CGI to create intricate visual effects, reducing the need for costly and time-consuming physical setups. Similarly, writers can utilize AI language models to draft initial ideas or structures, enabling them to focus on refinement and storytelling. By saving time and effort, these streamlined workflows empower creators to take on more ambitious projects without sacrificing quality or creativity.

The Risk of Over-Reliance on Technology

While technology offers remarkable benefits, over-reliance on it can stifle originality and depth. Automated tools, when used excessively, risk homogenizing creative outputs by relying on pre-existing patterns and trends. This can dilute the personal and emotional essence that sets human-driven creativity apart.

For example, AI-generated art, while visually captivating, may lead to an aesthetic uniformity that lacks the emotional nuance and individuality of human-made pieces. Similarly, writers who lean heavily on AI-generated text might find their unique voice overshadowed, resulting in content that feels impersonal or generic. Creators must remain mindful, ensuring that technology enhances their work rather than diminishing its distinctiveness.

Ethical Concerns and Bias

The adoption of technology in creative fields brings ethical challenges that cannot be overlooked. AI tools often rely on datasets that may carry biases, perpetuating stereotypes or excluding marginalized perspectives. Additionally, questions of intellectual property and authorship arise when algorithms generate content based on existing works.

For instance, AI trained on biased datasets might produce content that reinforces systemic inequities. Creators using such tools must actively address these biases to maintain fairness and inclusivity. Similarly, the use of AI-generated art or text raises questions about ownership: who truly "creates" the work, and how much credit belongs to the human creator versus the algorithm? Navigating these issues with integrity is essential for creators who wish to use technology responsibly.

Balancing Technology and Authenticity

The key to reaping the rewards of technological adoption while mitigating its risks lies in maintaining a balance between leveraging advanced tools and preserving the authenticity of the creative process. Technology should act as a collaborator rather than a replacement, enabling creators to enhance their work without compromising their unique voice or vision.

For example, a filmmaker might employ sophisticated visual effects to enrich their storytelling but anchor the narrative in emotionally resonant characters and themes. Similarly, a musician could integrate AI-generated sounds into their compositions while grounding the work in personal experiences and emotions. This balance ensures that technology serves as a complement to human creativity, not a substitute for it.

Navigating the Duality of Technology

Technological adoption in creative fields is neither inherently positive nor negative—it is a neutral force that can amplify or dilute creativity depending on how it is applied. By approaching technology with intentionality, creators can harness its capabilities to elevate their work while safeguarding the human elements that define authentic expression.

To navigate this duality effectively, creators should ask themselves critical questions: Does this tool align with my vision? Am I using it to enhance my process or replace it? How can I ensure my work remains personal and original in a technology-driven world? Addressing these questions thoughtfully allows creators to strike the right balance, using technology to enrich their art without compromising its essence.

A Forward-Looking Perspective

The risks and rewards of technological adoption underscore the importance of deliberate, thoughtful integration. Creators who embrace technology while remaining grounded in their creative vision can produce work that is both innovative and authentic. They demonstrate that advanced tools, when used responsibly, do not dilute originality but expand its scope.

By maintaining control over their creative processes and addressing ethical concerns with care, these creators chart a path forward that honors both tradition and progress. They prove that technology and authenticity can coexist, resulting in work that is impactful, meaningful, and deeply human.

Case Studies: Artists Harmonizing Old and New

The fusion of tradition and technology has given rise to a new wave of creators who bridge the divide between cultural heritage and contemporary innovation. These trailblazers prove that creativity thrives not as a choice between the old and the new, but as a harmonious blend of both. By finding synergy in this intersection, they craft art, fashion, and music that transcend temporal boundaries, resonating across cultures and audiences.

Takashi Murakami: Merging Nihonga with Digital Design

Japanese artist **Takashi Murakami** exemplifies the seamless integration of traditional artistry with modern techniques. Drawing inspiration from *Nihonga*, a classical Japanese painting style, Murakami employs digital tools to create vibrant, large-scale works that juxtapose intricate hand-painted details with bold, contemporary aesthetics. His signature motifs—like smiling flowers and anime-inspired characters—bridge the gap between Japanese cultural heritage and global pop culture.

Murakami's ability to blend traditional craftsmanship with digital innovation has garnered him worldwide acclaim. His works are featured in prestigious galleries, and his collaborations with global fashion brands have amplified his reach. By reinterpreting heritage through technology, Murakami demonstrates how the past can inform the future, resulting in creations that are both deeply rooted and refreshingly new.

Sabyasachi Mukherjee: Handcrafted Elegance Meets Modern Scaling

Indian fashion designer **Sabyasachi Mukherjee** is a master of preserving traditional craftsmanship while leveraging modern tools for scalability. Renowned for his exquisite use of handwoven textiles, intricate embroidery, and vintage aesthetics, Mukherjee combines these time-honored techniques with contemporary silhouettes to cater to a global audience.

To meet the demands of modern production without compromising authenticity, Mukherjee integrates technology into his creative process. Digital tools enable pattern replication and precision cutting, while the final embellishments remain meticulously handcrafted. This approach not only preserves Indian artisanal traditions but also sustains the livelihoods of artisan communities. Mukherjee's work underscores how technology can support, rather than replace, cultural heritage.

Björk: Innovating Music with Tradition and Technology

Icelandic artist **Björk** is a pioneer in blending natural elements with digital innovation in her music. Her compositions juxtapose electronic beats and effects with traditional Icelandic instruments and sounds from the natural world, such as bird calls and the rhythm of waterfalls. The result is a sonic experience that feels both futuristic and deeply organic.

By incorporating tools like AI and digital soundscapes, Björk expands the boundaries of musical storytelling while maintaining a profound connection to her roots. Her work exemplifies how technology, when used thoughtfully, can enhance rather than dilute cultural and emotional authenticity.

Refik Anadol: Digital Art Rooted in Memory and Culture

Turkish-born artist **Refik Anadol** uses artificial intelligence and big data to create immersive digital art installations that explore memory, culture, and technology. Anadol's work analyzes massive datasets, such as historical archives and cultural records, to generate dynamic visualizations that bring abstract concepts to life.

His installations often reimagine cultural themes, such as traditional architecture or significant historical events, through a digital lens. By blurring the lines between art, science, and technology, Anadol demonstrates how modern tools can preserve and reinterpret cultural narratives in innovative and engaging ways.

Other Inspiring Examples of Heritage and Innovation

✓ **Ai Weiwei**, the Chinese artist and activist, combines traditional materials like porcelain and wood with digital tools for conceptual planning and execution. His works critique societal issues while celebrating the value of craftsmanship.

✓ **Christian Louboutin**, the iconic shoe designer, merges traditional handcrafting techniques with 3D modeling, producing footwear that balances artisanal quality with cutting-edge design.

✓ **Marina Abramović**, a pioneer of performance art, explores virtual reality to reinterpret her deeply human works, blending the emotional depth of tradition with the immersive possibilities of modern technology.

Lessons from the Intersection of Old and New

These case studies illustrate that the fusion of tradition and technology is not about choosing one over the other but about enhancing and amplifying the strengths of both. Creators like Murakami, Mukherjee, Björk, and others show that by embracing both heritage and innovation, it is possible to produce work that feels timeless yet forward-thinking.

This approach offers a roadmap for modern creators, demonstrating how respect for history can coexist with bold experimentation. By harmonizing the authenticity of tradition with the possibilities of technology, these artists craft works that resonate deeply with audiences across cultures and generations.

A Blueprint for Meaningful Innovation

The success of these creators highlights the transformative potential of blending the old with the new. Their works serve as a reminder that innovation rooted in heritage carries a unique depth and authenticity. By honoring cultural legacies while embracing modern tools, they set an inspiring example for future generations, proving that creativity is at its most powerful when it bridges time, culture, and technology.

Shaping the Future of Creative Innovation

The integration of technology into creative fields offers unprecedented opportunities to rethink how we conceptualize, create, and connect. However, this evolution also presents challenges, requiring creators to adopt technological advancements with intention and mindfulness. While modern tools can amplify creative potential and unlock new possibilities, they also pose risks to originality and authenticity. Striking a balance between innovation and authenticity will define a future where creativity thrives in both depth and breadth.

Harnessing the Potential of Technology

Technology provides creators with tools that enhance efficiency, broaden artistic horizons, and facilitate global collaboration. From AI-assisted design platforms to virtual and augmented reality, these innovations empower creators to visualize and execute ideas with precision and scale. By leveraging such advancements, creators can transcend traditional constraints and explore bold, uncharted concepts.

For instance, architects can now use generative design software to optimize sustainable structures for functionality and aesthetics. Similarly, filmmakers can craft immersive worlds that merge physical and digital elements through augmented reality. These tools encourage creators to experiment and innovate, fostering a culture of exploration that drives progress across disciplines.

The Responsibility of Creators

With technological power comes the responsibility to wield it thoughtfully. The speed and convenience of modern tools can tempt creators to prioritize efficiency over depth, risking work that feels generic or impersonal. Maintaining authenticity in this context requires creators to approach technology as a collaborator that enhances their vision rather than a shortcut that bypasses the creative process.

For example, an artist might use AI to generate foundational concepts but refine the final piece to reflect their unique perspective and emotional depth. Similarly, a musician could explore AI-generated melodies but rely on live performances to infuse their work with human nuance. This intentional approach ensures that technology complements rather than dominates the creative process.

Navigating Ethical and Cultural Challenges

The adoption of technology in creative fields raises significant ethical and cultural considerations. Issues such as data privacy, algorithmic bias, and questions of ownership and authorship require proactive strategies to prevent unintended consequences. Creators must navigate these challenges responsibly to maintain integrity and build trust with their audiences.

For example, AI-generated content prompts discussions about originality and intellectual property. Defining clear boundaries for how technology is used can help creators maintain the authenticity of their work. Similarly, ensuring that datasets used in AI tools are inclusive and free from bias prevents the perpetuation of stereotypes. By addressing these concerns thoughtfully, creators can contribute positively to the broader cultural and ethical landscape.

Balancing Tradition and Modernity

While technology enables unprecedented innovation, it is essential not to lose sight of the enduring value of tradition. Many celebrated works draw strength from their cultural roots, blending timeless techniques with contemporary ideas. The future of creative innovation lies in this balance between old and new, where heritage informs progress, and technology amplifies timeless values.

For instance, textile artisans using digital tools to design traditional motifs ensure their cultural legacy remains relevant in modern contexts. Similarly, authors who incorporate ancient storytelling techniques into contemporary narratives create works that resonate across generations. This integration ensures that creativity remains both grounded in heritage and forward-thinking.

Fostering Interdisciplinary Collaboration

The future of creative innovation will increasingly be shaped by interdisciplinary collaboration, where creators from diverse fields work together to share knowledge and perspectives. Technology acts as a bridge, enabling artists, scientists, engineers, and thinkers to collaborate on projects that transcend traditional boundaries. This cross-pollination results in richer, more impactful creations that reflect the complexities of the modern world.

For example, collaborations between artists and technologists have produced interactive installations that blend art and science, offering new ways to engage audiences. Similarly, partnerships between designers and engineers have created innovative products that marry aesthetic appeal with functional excellence. These examples highlight the transformative potential of interdisciplinary teamwork in shaping the future of creativity.

Empowering the Next Generation of Creators

To sustain innovation, it is critical to empower the next generation of creators with the skills, tools, and ethical awareness needed to navigate the digital age effectively. Education and mentorship play vital roles in fostering a culture of curiosity, adaptability, and resilience among aspiring artists, writers, and designers.

Programs that teach young creators how to use technology responsibly while exploring traditional techniques provide a balanced foundation for artistic growth. Ensuring access to diverse creative tools and resources also promotes inclusivity, breaking down geographic and socioeconomic barriers to innovation. Investing in the next generation is not only an investment in their futures but in the future of creativity itself.

Redefining the Creative Process

Technology is fundamentally redefining the creative process, transforming how ideas are conceived, developed, and shared. This evolution challenges traditional notions of authorship and originality, encouraging creators to embrace new approaches while staying true to their core vision. Adopting a growth mindset allows creators to view technological change not as a threat but as an opportunity to expand their craft.

Adaptability is key. Creators must remain open to experimentation while grounding their work in purpose and identity. By redefining the creative process in ways that prioritize authenticity, collaboration, and innovation, they can shape a future where technology enhances creativity rather than diminishing its human essence.

A Future Built on Thoughtful Innovation

The future of creative innovation will be shaped by the choices we make today. By approaching technology with intention and awareness, creators have the power to redefine what it means to be creative in the digital age. Thoughtful integration of tools and techniques ensures that creativity remains a deeply human endeavor, enriched by technology but never overshadowed by it.

As we navigate this transformative era, the onus lies with creators to balance the possibilities of innovation with the timeless values of authenticity and tradition. By doing so, they can ensure that their work remains meaningful, impactful, and uniquely their own, laying the foundation for a future where creativity knows no bounds.

Case Studies: Artists Harmonizing Old and New

In today's creative landscape, certain artists have skillfully bridged the divide between tradition and technology, demonstrating how these forces can complement rather than conflict. By honoring their cultural roots while embracing modern tools, these creators craft works that resonate across time, combining the depth of heritage with the possibilities of innovation. Their journeys highlight the transformative potential of balancing wisdom from the past with the advancements of the future.

Takashi Murakami: Redefining Nihonga with Digital Precision

Japanese artist Takashi Murakami seamlessly merges the classical Nihonga painting style with contemporary digital tools. Nihonga, known for its meticulous detail and natural pigments, is a cornerstone of Japanese art. Murakami revitalizes this tradition by incorporating digital technology, creating vibrant, large-scale works that blend intricate hand-painted elements with bold, pop culture aesthetics.

His signature "superflat" style combines traditional motifs with anime-inspired characters and psychedelic patterns. Through digital platforms, Murakami achieves precision, scalability, and global accessibility, while staying true to his cultural heritage. Collaborations with luxury brands like Louis Vuitton and exhibitions in leading galleries showcase how tradition can coexist with cutting-edge innovation, captivating diverse audiences.

Sabyasachi Mukherjee: Marrying Artisanal Heritage with Technology

Indian fashion designer Sabyasachi Mukherjee exemplifies how traditional craftsmanship can thrive in the digital age. Known for intricate hand-embroidered textiles and vintage aesthetics, Mukherjee integrates technology to streamline production without compromising authenticity.

Digital tools allow him to experiment with patterns and silhouettes during the design phase, ensuring innovation while preserving the handmade details that define his work. This approach not only elevates Indian couture on a global stage but also supports artisanal communities, keeping their skills relevant. Mukherjee's work demonstrates that technology, when used thoughtfully, can act as a custodian of tradition rather than a threat.

Björk: Weaving Nature and Technology into Music

Icelandic artist Björk is a trailblazer in blending traditional elements with futuristic innovation. Her music combines electronic beats, natural sounds, and traditional Icelandic instruments, creating immersive sonic landscapes that are both deeply organic and technologically advanced.

Using bespoke instruments like digitally enhanced gamelans and MIDI-driven harps, Björk fuses the sounds of nature—bird calls, wind, and waterfalls—with synthetic rhythms. This approach allows her to push musical boundaries while staying rooted in her heritage. Her work highlights how tradition can inspire technological experimentation, resulting in music that feels timeless yet ahead of its time.

Refik Anadol: Preserving Cultural Memory with AI

Turkish artist Refik Anadol uses artificial intelligence and big data to reinterpret cultural narratives. His digital installations transform historical archives and cultural records into dynamic visual experiences, bridging the gap between memory and modernity.

In his project *Melting Memories*, Anadol visualizes brain wave activity associated with memory, blending art and neuroscience. By digitizing and reimagining cultural artifacts, he ensures their relevance in the digital age. Anadol's work exemplifies how AI can preserve heritage while exploring new creative frontiers, making cultural memory accessible to a global audience.

Christian Louboutin: Craftsmanship Enhanced by Technology

Renowned shoe designer Christian Louboutin combines traditional artistry with modern precision. Famous for his iconic red-soled shoes, Louboutin integrates handcrafting techniques with advanced technologies like 3D modeling and laser engraving.

This synergy allows him to explore innovative designs while maintaining the artisanal quality of his brand. By embracing both tradition and technology, Louboutin creates footwear that balances timeless luxury with contemporary appeal.

Marina Abramović: Transforming Performance Art with Digital Tools

Performance artist Marina Abramović has embraced technology to reinterpret her emotionally charged works. Known for live performances exploring endurance and vulnerability, Abramović now uses virtual reality to expand her artistic reach.

Her VR piece *Rising* immerses viewers in a digitally constructed environment addressing climate change, eliciting visceral emotional responses. By combining traditional performance art with digital tools, Abramović maintains her focus on human connection while exploring new dimensions of engagement.

Lessons in Harmonizing Old and New

These case studies demonstrate that integrating tradition and technology is not about compromise but about enhancement. Artists like Murakami,

Mukherjee, Björk, and others prove that innovation can coexist with heritage, resulting in creations that feel both timeless and forward-thinking. When tradition informs innovation and technology amplifies authenticity, the outcome is work that resonates deeply across cultures and generations.

By embracing this balance, creators ensure their work honors the richness of the past while harnessing the possibilities of the future, crafting a legacy that seamlessly bridges both worlds. This harmonious fusion serves as a model for creative industries, illustrating the power of blending heritage with modernity to shape meaningful, impactful art.

A Path Forward: Balancing Tradition and Technology

The intersection of tradition and technology offers one of the most profound opportunities for modern creators. By navigating this space with intention, creators can honor the timeless wisdom of heritage while harnessing the transformative potential of innovation. This approach not only enriches the creative process but ensures that art, design, and craftsmanship remain relevant, impactful, and resonant in an ever-evolving world.

Honoring the Past

Heritage provides creators with a deep wellspring of inspiration, offering techniques, stories, and philosophies refined through generations. These traditions anchor contemporary work, providing a sense of continuity and authenticity that transcends trends. By preserving and respecting these practices, creators imbue their work with depth and meaning derived from their connection to something larger than themselves.

For instance, artisans using traditional weaving techniques create textiles infused with cultural history and identity that machines cannot replicate. Similarly, writers who draw from ancient myths or storytelling structures bring universal human truths into contemporary narratives. Honoring the past doesn't mean resisting change; it means using tradition as a foundation to build something new and enduring.

Embracing Innovation

Innovation propels creativity forward, opening doors to possibilities that were once unimaginable. Modern tools and technologies enable creators to experiment, iterate, and expand their horizons. By embracing innovation, creators can push boundaries, explore uncharted territories, and engage with broader audiences.

For example, digital platforms allow visual artists to share their work globally, transcending geographical barriers. Generative AI and machine learning provide musicians and filmmakers with tools to create unique compositions and immersive experiences. Innovation offers opportunities for growth, adaptation, and reinvention, allowing creators to stay relevant while remaining true to their vision.

Thoughtful Engagement: The Key to Balance

Balancing tradition and technology requires thoughtful engagement. Creators must approach their work with intention, evaluating how and why they incorporate modern tools. This involves questioning whether technology enhances or detracts from their creative vision and ensuring that their use of innovation aligns with their core values.

For example, architects might use generative design software to optimize functionality while incorporating traditional materials to preserve cultural significance. Similarly, filmmakers might employ CGI for visual effects but ground their narratives in authentic human emotions. Thoughtful engagement ensures that technology acts as an ally, amplifying rather than overshadowing the creator's voice.

The Courage to Experiment

Experimentation is at the heart of blending tradition and technology. Creators must be willing to challenge conventions, take risks, and combine elements in unexpected ways. This boldness leads to discoveries that redefine norms and expand creative horizons.

Designers who merge traditional craftsmanship with 3D printing, for instance, create pieces that are both innovative and rooted in history. Musicians blending ancient instruments with electronic beats honor their heritage while pushing the boundaries of their genres. Fearless experimentation allows creators to unite the best of both worlds, creating works that are both distinctive and transformative.

Preserving Authenticity and Depth

Authenticity and depth are the cornerstones of meaningful creativity. Regardless of the tools or techniques employed, creators must stay true to their vision, ensuring their work reflects their unique perspective and values. Authenticity fosters connections that transcend time and technology, resonating deeply with audiences.

To preserve authenticity, creators can integrate technology in ways that complement the human touch. Whether it's the texture of a hand-painted mural, the emotional nuance of a live performance, or the cultural richness of traditional storytelling, these elements must remain central to the creative process. Authenticity ensures that innovation enhances rather than diminishes the depth and impact of the work.

Shaping the Future of Creativity

Balancing tradition and technology is about more than individual works; it is about shaping the future of creativity itself. By demonstrating how these forces can coexist, creators set a powerful example for future generations. This balance ensures cultural heritage remains vibrant in a digital world while fostering inclusivity, allowing diverse voices and traditions to thrive on the global stage.

Educational programs that teach young creators to value both tradition and innovation play a crucial role in this process. By equipping them with the skills to navigate modern complexities while respecting their cultural roots, we cultivate a generation of creators who are both grounded and visionary.

A Legacy of Harmony

Balancing tradition and technology is not merely a challenge—it is an opportunity to create a legacy of harmony. By embracing the richness of their heritage while exploring the potential of innovation, creators produce works that are timeless and forward-thinking. This approach ensures that creativity remains a powerful force for connection, growth, and transformation.

Through thoughtful engagement, fearless experimentation, and unwavering authenticity, creators can forge a path that honors the past while shaping the future. This path enriches their work and inspires others, proving that the interplay between tradition and technology is not a compromise but a symphony of possibilities.

6.2 Overcoming Societal and Self-Imposed Restrictions

The creative process, while liberating, is often shaped and constrained by external societal expectations and internal personal barriers. Overcoming these restrictions is essential for creators to unlock their full potential and produce meaningful, transformative work. By challenging cultural norms, addressing self-doubt, breaking through creative blocks, and fostering inclusivity, creators can transcend limitations and embrace innovation.

Challenging Cultural Norms in Creative Expression

Cultural norms serve as both a foundation and a boundary for creativity. While they provide a shared framework of values and aesthetics, they can also impose restrictions on what is considered acceptable or valuable in artistic expression. These norms, shaped by tradition, societal expectations, and industry standards, often create invisible barriers that discourage creators from exploring unconventional or controversial territories. Overcoming these barriers demands boldness, authenticity, and a willingness to question the status quo.

The Weight of Tradition and Expectation

Tradition is a double-edged sword. On one hand, it preserves heritage and identity, offering creators a rich repository of techniques, themes, and symbols. On the other, it can stifle innovation, as societal expectations often pressure creators to conform to established norms. In many cultures, deviation from traditional forms is seen as disrespectful or unnecessary, creating tension for artists seeking to explore new ideas.

For instance, a visual artist in a conservative society might face backlash for using provocative imagery to address social injustices. Similarly, a writer discussing taboo topics like LGBTQ+ rights, mental health, or political corruption may encounter censorship or ostracism. Such constraints not only limit individual creators but also inhibit the evolution of artistic and cultural expression.

Breaking Through Taboo Subjects

Addressing taboo subjects is one of the most challenging yet impactful ways to challenge cultural norms. Topics considered sensitive or controversial often remain unexplored, leaving significant aspects of the human experience in the shadows. However, some of history's most transformative works have come from creators who dared to confront these issues.

For example, filmmakers like Mira Nair and Asghar Farhadi have explored themes of gender, class, and family dynamics within deeply traditional societies. Their work pushes artistic boundaries while inviting audiences to critically examine their beliefs. Similarly, writers like Chimamanda Ngozi Adichie and Salman Rushdie use their narratives to challenge cultural and political norms, broadening perspectives and sparking vital conversations.

Criticism, Censorship, and Resilience

Creators challenging cultural norms often face criticism and censorship from gatekeepers of tradition. Their work may be dismissed, suppressed, or even banned. Yet, such opposition underscores the significance of their contributions. Resistance is often a sign that their work is provoking critical thought and pushing boundaries.

For instance, controversial films, books, or performances frequently gain increased public attention when censored, amplifying their message and sparking widespread dialogue. This phenomenon highlights the power of art to challenge entrenched beliefs and inspire change. Creators who persist despite adversity pave the way for future generations to express themselves more freely.

The Role of Authenticity in Challenging Norms

Authenticity lies at the heart of challenging cultural norms. It reflects a creator's commitment to staying true to their vision and values, even in the face of resistance. Authentic creators do not provoke for provocation's sake; they use their work to explore genuine questions and share unique perspectives. This sincerity often resonates with audiences, fostering empathy and understanding.

For example, visual artist Ai Weiwei critiques political oppression and advocates for human rights through his deeply personal and authentic works. His art, rooted in his lived experiences, carries credibility and impact, making it a powerful tool for social change. Similarly, authors like Arundhati Roy intertwine personal narratives with broader critiques of injustice, ensuring their work remains both intimate and universal.

Expanding the Scope of Possibilities

By challenging cultural norms, creators expand the boundaries of what is possible in their fields. Their willingness to question conventions and introduce new ideas inspires others to think differently and explore uncharted territories. This ripple effect fosters a more inclusive and dynamic creative landscape where diverse voices and perspectives can thrive.

For instance, feminist literature in the 20th century redefined storytelling by centering women's experiences and challenging patriarchal narratives. Contemporary artists blending traditional forms with modern techniques demonstrate that cultural heritage and innovation can coexist harmoniously. These trailblazers act as catalysts for change, encouraging others to question assumptions and embrace new possibilities.

The Path Forward: Courage and Connection

Challenging cultural norms requires both courage and connection. Creators need the bravery to confront resistance and the sensitivity to engage their audiences in meaningful dialogue. By fostering empathy and understanding, they can bridge divides and inspire change without alienating those who might initially oppose their ideas.

In an increasingly interconnected yet divided world, challenging cultural norms is not just an artistic endeavor—it is an essential act of leadership. Creators who dare to question conventions and expand the boundaries of expression contribute to a more inclusive and equitable society. Their work exemplifies the transformative power of art in shaping our collective future, proving that creativity is not just a reflection of culture but a force for its evolution.

Navigating Self-Doubt and the Fear of Failure

Self-doubt and the fear of failure are universal challenges that can stifle creative expression, often manifesting as internal barriers that limit experimentation, risk-taking, and the willingness to share one's work.

Overcoming these hurdles requires self-awareness, resilience, and intentional strategies to unlock creativity's full potential.

The Paralysis of Self-Doubt

Self-doubt can paralyze creators, causing them to question the value and quality of their work. A writer may hesitate to explore bold themes, fearing negative reception, while a musician might avoid performing new compositions out of uncertainty about their impact. This hesitation, often fueled by a fear of judgment, criticism, or unfavorable comparisons, creates a cycle that undermines confidence and growth.

Creators plagued by self-doubt frequently abandon projects or never start them. Each unfulfilled idea reinforces feelings of inadequacy, amplifying the voice of doubt. Breaking free from this cycle begins with recognizing its existence and reclaiming the joy of creating without fear.

Reframing Failure as Growth

The fear of failure can be diminished by reframing it as an essential component of progress. Failure is not a sign of inadequacy but a natural outcome of striving for innovation and growth. Each setback provides valuable lessons, offering insights into what works and what doesn't, ultimately refining creative efforts.

Celebrated creators like Thomas Edison and J.K. Rowling have often highlighted how failure shaped their successes. Edison's perspective—"I have not failed. I've just found 10,000 ways that won't work"—exemplifies the transformative power of viewing failure as a stepping stone rather than a stopping point. This mindset reduces failure's emotional impact, fostering perseverance and resilience.

Practical Strategies to Overcome Self-Doubt

Creators can employ actionable strategies to navigate self-doubt and mitigate the fear of failure:

✓ **Journaling:** Writing about fears and uncertainties clarifies thoughts and reveals the root causes of doubt. This reflective practice fosters self-awareness and helps generate solutions.

✓ **Mindfulness:** Techniques like meditation or deep breathing can manage anxiety and keep creators grounded in the present, reducing the mental noise that feeds self-doubt.

✓ **Constructive Feedback:** Seeking input from trusted peers or mentors offers fresh perspectives and highlights strengths, while positive reinforcement builds confidence.

✓ **Setting Realistic Goals:** Breaking projects into smaller, manageable milestones reduces overwhelm and creates a sense of accomplishment at each step.

✓ **Celebrating Progress:** Recognizing even small victories reinforces self-belief and fosters a positive association with the creative process.

These strategies help shift the focus from unattainable perfection to steady progress, promoting action over hesitation and building confidence through experience.

The Role of Supportive Networks

A strong support system can be transformative in overcoming self-doubt. Surrounding oneself with encouraging peers, mentors, or creative communities fosters validation and belonging. Sharing ideas, receiving feedback, and learning from others' experiences can reignite motivation and instill the courage to take risks.

Creative workshops, artist collectives, and online forums provide safe spaces for collaboration and growth. These environments remind creators that self-doubt is a shared experience and that overcoming it is possible through mutual support and shared learning.

Focusing on the Journey Over the Outcome

A liberating approach to navigating self-doubt and the fear of failure is prioritizing the creative process over the final product. By valuing exploration, experimentation, and learning, creators can detach from external validation and rediscover the intrinsic joy of creating.

For example, a painter might explore abstract forms purely for the experience of discovery, or a musician might compose without concern for commercial success. This mindset encourages fearless experimentation, often leading to unexpected breakthroughs and deeper personal fulfillment.

Building Resilience Through Practice

Overcoming self-doubt and fear of failure is an ongoing practice. Each creative endeavor, regardless of its success, builds resilience and reinforces the belief that challenges can be overcome. Consistently facing fears and pushing through discomfort strengthens creators' ability to navigate uncertainty and fosters a deeper connection to their craft.

As creators grow attuned to their strengths and limitations, they develop a greater appreciation for their unique contributions. This self-assurance becomes a powerful antidote to doubt, empowering them to pursue ambitious projects with confidence and clarity.

A Path Toward Fearless Creativity

Self-doubt and the fear of failure are natural parts of the creative journey but need not define it. By reframing failure, adopting intentional strategies, and cultivating supportive networks, creators can transcend these challenges and unlock their true potential. The process of overcoming these barriers not only enhances creativity but also builds resilience, equipping creators to face future obstacles with determination and purpose.

Fearless creativity lies in embracing uncertainty and trusting in the value of one's unique voice. By moving beyond self-imposed restrictions, creators open the door to boundless possibilities, crafting work that is both authentic and impactful.

Breaking Free from Creative Blocks and Limitations

Creative blocks are a natural, albeit frustrating, part of the creative process. Often arising from a mix of external pressures and internal struggles, they can leave creators feeling disconnected and stagnant. However, these moments are not just obstacles but opportunities for reflection, reinvention, and growth.

Identifying the Roots of Creative Blocks

The first step in overcoming a creative block is identifying its cause. External factors like deadlines, financial stress, or negative feedback can create tension that hinders creativity. Internally, self-doubt, perfectionism, or a lack of direction can lead to paralysis. Understanding these triggers allows creators to address the problem at its source.

For instance, a writer struggling to move forward might discover that fear of criticism is the real barrier. Similarly, an artist feeling uninspired might recognize that monotony in their routine is dulling their creativity. By pinpointing these underlying issues, creators can take meaningful steps to reignite their momentum.

Changing Perspectives to Unlock Creativity

Sometimes, breaking free from a block requires shifting one's perspective. Stepping away from a problem can provide clarity, allowing fresh ideas to surface. A walk, meditation session, or engagement in an unrelated activity can reset the mind and inspire new approaches.

Experimentation is another powerful tool. Exploring new techniques, styles, or mediums can refresh creative energy. For example, a painter could try digital art, or a musician might experiment with a new instrument or genre. These explorations often lead to unexpected inspiration and new skills.

Collaboration can also open up new possibilities. Partnering with others exposes creators to fresh ideas and perspectives, breaking the isolation that often accompanies a block. For instance, a filmmaker might work with a visual artist to gain new storytelling insights, or a writer might join a workshop to exchange ideas and receive constructive feedback.

Breaking Down Tasks into Manageable Goals

The enormity of a creative project can feel overwhelming during a block. Breaking tasks into smaller, achievable steps can reduce pressure and create a sense of progress. Focusing on a single paragraph, sketch, or melody at a time helps rebuild confidence.

Celebrating small victories is equally essential. Recognizing even minor achievements reinforces positive momentum and encourages continued effort. Over time, these incremental steps lead to significant breakthroughs, reigniting creativity and restoring motivation.

Rediscovering Joy Through Play and Curiosity

Creative blocks often arise when creators lose sight of the joy that drew them to their craft. Reconnecting with that initial passion through playful, curiosity-driven activities can reignite inspiration.

For example, a writer might free-write without worrying about structure, simply letting ideas flow. An artist might create quick, abstract doodles with no specific goal. These exercises strip away the weight of expectations and allow creators to rediscover the pleasure of the creative process.

Establishing Routines to Nurture Creativity

Consistency can help prevent and overcome creative blocks. Establishing a routine that prioritizes creative time fosters discipline and minimizes distractions. Even a daily commitment of 15 minutes to creative work can build momentum and make creativity feel more accessible.

Incorporating activities that inspire and nurture creativity, such as journaling, reading, or exploring new ideas, ensures a steady flow of inspiration. A structured yet flexible routine provides both stability and room for exploration.

Embracing Imperfection as Part of the Process

Fear of imperfection often fuels creative blocks, making creators hesitant to start or continue their work. Embracing imperfection as an integral part of creativity can liberate them from this fear.

Many renowned creators emphasize the importance of producing imperfect drafts or sketches as stepping stones to brilliance. Acknowledging that the first attempt doesn't need to be perfect allows creators to take risks, experiment freely, and ultimately refine their work.

Transforming Challenges into Opportunities

Creative blocks can feel like barriers, but they often present opportunities for growth. These moments encourage creators to pause, reassess, and explore new directions. Pushing through limitations often leads to unexpected discoveries and breakthroughs.

For instance, an architect might use a creative block as an opportunity to study innovative designs or draw inspiration from nature. A musician might revisit unfinished pieces, finding fresh ideas in older work. These challenges, when approached with curiosity and openness, often lead to rewarding outcomes.

Adopting a Growth-Oriented Mindset

Breaking free from creative blocks is not about eliminating them entirely but developing a mindset that embraces the ebb and flow of creativity. Viewing blocks as temporary and as part of the creative journey reduces their emotional weight and fosters resilience.

By focusing on progress over perfection and treating each challenge as a learning opportunity, creators can cultivate a mindset that turns stagnation into growth. Each small step forward builds confidence, laying the foundation for future success.

A Journey of Resilience and Discovery

Creative blocks are not failures but opportunities to strengthen one's craft and character. By identifying triggers, experimenting with new approaches, and celebrating small wins, creators can navigate these challenges with resilience and grace.

Ultimately, overcoming creative blocks requires embracing the uncertainties of the journey while trusting in the process. With patience and persistence, creators can transform moments of stagnation into stepping stones, emerging with renewed energy, inspiration, and confidence to push the boundaries of their craft.

Embracing Diversity and Inclusivity in Modern Creativity

Diversity and inclusivity are vital pillars of a thriving creative ecosystem. They enrich artistic expression, challenge entrenched biases, and foster deeper connections between creators and audiences. By embracing diverse perspectives and amplifying underrepresented voices, creators can push the boundaries of their craft and produce work that authentically reflects the multifaceted nature of human experience. Achieving true inclusivity, however, demands intentionality and a concerted effort to dismantle societal and structural barriers that hinder creative expression.

The Power of Diversity in Creativity

Diversity brings a wealth of perspectives, traditions, and lived experiences to the creative process, enhancing its depth and impact. By drawing from various cultural, socioeconomic, and personal backgrounds, creators can explore themes that resonate universally while celebrating the uniqueness of individual stories. This diversity strengthens emotional and intellectual engagement, allowing creative work to transcend geographical and cultural boundaries.

For instance, authors like Chinua Achebe and Chimamanda Ngozi Adichie have brought African experiences and histories to global audiences, challenging stereotypes and broadening perspectives. Similarly, filmmakers such as Ava DuVernay and Bong Joon-ho have redefined storytelling by centering marginalized voices and addressing critical social issues. These creators demonstrate how diversity transforms creative landscapes, making them more reflective of the complexities of modern society.

Challenging Biases to Broaden Perspectives

Despite its potential, the creative industry often mirrors the biases of the societies in which it operates. These biases can marginalize certain voices, perpetuate stereotypes, and limit innovation. Historical underrepresentation of women, LGBTQ+ individuals, and people of color in creative fields has reinforced dominant narratives at the expense of inclusivity.

Creators who challenge these biases play a crucial role in reshaping the cultural landscape. By casting actors from diverse backgrounds, exploring unconventional narratives, or depicting underrepresented communities with authenticity, they foster greater empathy and understanding. These efforts not only enrich creative expression but also dismantle systemic inequities, paving the way for a more inclusive artistic ecosystem.

Amplifying Underrepresented Voices

Inclusivity extends beyond representation—it requires actively amplifying voices that have been historically excluded. Providing platforms for underrepresented creators brings fresh perspectives into the mainstream, challenging traditional norms and encouraging innovation.

Initiatives like the Sundance Film Festival and the Women's Prize for Fiction highlight works by creators from diverse backgrounds, offering visibility to stories that might otherwise be overlooked. These platforms empower creators to explore unconventional themes and formats, ensuring that the creative landscape reflects a broader spectrum of human experiences.

Addressing Unconscious Bias in Creativity

Unconscious biases can subtly shape creative decisions, influencing casting, storytelling, or design in ways that perpetuate stereotypes.

Acknowledging and addressing these biases is essential for fostering authenticity and inclusivity in creative work.

For example, a filmmaker portraying a culture different from their own must engage in thoughtful research and consultation with members of that community to ensure respectful representation. Similarly, product designers must consider accessibility and cultural nuances to create inclusive designs. By being mindful of these biases, creators can build trust with their audiences and produce work that resonates with integrity and inclusivity.

The Role of Collaboration in Inclusivity

Collaboration is a powerful avenue for fostering inclusivity. Working with individuals from diverse backgrounds exposes creators to new perspectives, challenging assumptions and enriching the creative process. Collaborative projects often result in innovative outcomes that reflect the collective wisdom of a diverse team.

For instance, cross-cultural collaborations in music, such as those by the Silkroad Ensemble, blend traditional and contemporary styles to create globally resonant works. Similarly, collaborative storytelling in film or theater brings together voices from varied experiences, producing narratives that feel authentic and relatable. These partnerships demonstrate the transformative power of inclusivity in breaking down cultural barriers and fostering global understanding.

Fostering Empathy Through Inclusive Storytelling

Inclusive creativity has the unique ability to build empathy by presenting audiences with stories and experiences that differ from their own. By highlighting diverse perspectives, creators challenge assumptions, provoke thought, and foster a deeper understanding of shared humanity.

Films like *Moonlight* and *Parasite* have achieved universal acclaim by portraying authentic, nuanced experiences that resonate across cultures. Such works encourage viewers to step into others' shoes, breaking down prejudices and fostering connections. This empathy-building aspect of inclusive creativity is essential for bridging divides in an increasingly polarized world.

Building a Path Toward Authentic Inclusivity

Diversity and inclusivity in creativity are not trends—they are necessities for meaningful and impactful artistic expression. By embracing these principles, creators ensure that their work reflects the richness and complexity of the human experience while challenging societal norms and biases.

True inclusivity requires more than representation; it demands a commitment to authenticity, collaboration, and continuous learning. It involves creating spaces where underrepresented voices are heard, celebrated, and empowered to thrive. This approach not only enriches the creative process but also contributes to a more equitable cultural landscape.

Creativity as a Catalyst for Change

Inclusive creativity serves as a bridge between cultures, perspectives, and identities, reminding us of our shared humanity while celebrating our differences. It has the power to inspire change by challenging dominant narratives, amplifying marginalized voices, and fostering dialogue.

In a world that is both globally connected and deeply divided, the role of inclusive creativity cannot be overstated. By committing to diversity and inclusivity, creators can craft works that resonate universally, foster understanding, and contribute to a more just and inclusive society.

Strategies for Empowering Personal and Collective Innovation

To overcome societal and self-imposed limitations, creators must adopt strategies that foster both individual growth and collaborative potential. These approaches encourage experimentation, learning, and purpose-driven work, unlocking new dimensions of creativity and impact. By combining personal drive with collective synergy, creators can transcend barriers and achieve transformative innovation.

Collaborative Creativity: Unlocking New Possibilities

Collaboration is a catalyst for innovation, offering creators access to diverse perspectives, skills, and experiences. Working with others fosters a dynamic exchange of ideas, encouraging participants to venture beyond their comfort zones and explore uncharted territories.

For example, cross-disciplinary collaborations, such as those between artists and engineers or scientists and designers, often lead to groundbreaking projects that expand the boundaries of both fields. In fashion, partnerships between traditional craftsmen and contemporary designers have revived heritage techniques, introducing them to modern audiences. These examples demonstrate how collaboration amplifies creativity and strengthens the collective impact of creative endeavors.

Continuous Learning: Adapting to an Evolving World

Creativity thrives on curiosity and adaptability. In a rapidly evolving landscape, continuous learning helps creators expand their skills, embrace emerging tools, and stay inspired. This commitment to growth ensures relevance and fosters innovation.

Workshops, mentorship programs, and online courses offer valuable opportunities for learning. For instance, a filmmaker might explore cutting-edge editing software, while a writer delves into digital publishing platforms or emerging genres. Conferences and creative communities also provide fertile ground for shared learning, where creators can exchange ideas and gain fresh insights.

By embracing lifelong learning, creators remain agile and ready to navigate both challenges and opportunities in the modern creative environment.

Purpose-Driven Creation: Aligning Work with Values

Purpose-driven creation fuels motivation and lends authenticity to the creative process. By aligning their work with personal values or societal impact, creators tap into a deeper sense of fulfillment and meaning.

For instance, an artist might focus on themes of environmental sustainability, while a filmmaker amplifies marginalized voices through storytelling. These creators channel their passion into works that resonate deeply with audiences, inspiring meaningful change.

Purpose-driven creation also fosters resilience, helping creators stay focused and persevere through criticism or setbacks by anchoring their efforts in a larger mission.

Safe Spaces for Experimentation: Encouraging Risk-Taking

Innovation often requires stepping into the unknown. Creating safe spaces for experimentation—where failure is seen as a learning opportunity—encourages creators to take risks and explore new possibilities without fear of judgment.

These spaces can take various forms, such as physical studios for prototyping or online communities where creators share in-progress work and exchange feedback. Supportive peer groups and mentorship programs also provide environments where creators feel empowered to experiment.

For example, companies use hackathons to foster creativity, providing employees with low-pressure settings to test ideas. Similarly, artists' residencies offer dedicated time and space for exploration, nurturing growth over perfection. These environments inspire creators to push boundaries and innovate boldly.

Mindset Shifts: Turning Constraints into Opportunities

A key strategy for empowering innovation is reframing restrictions as opportunities. This mindset shift transforms challenges into catalysts for creativity, fostering resilience and adaptability.

For example, filmmakers working with limited budgets often develop inventive ways to tell compelling stories, leveraging resourcefulness to create impactful visuals. Writers constrained by word counts or formats, such as poetry or flash fiction, produce works of precision and power by focusing on clarity and economy.

By embracing constraints as opportunities, creators enhance their ability to adapt and discover unexpected solutions, turning limitations into stepping stones for growth.

Synergy Between Personal and Collective Innovation

Personal and collective innovation are deeply interconnected. By fostering collaboration, committing to continuous learning, and aligning work with purpose, creators not only empower themselves but also contribute to a more dynamic creative ecosystem. Safe spaces for experimentation and a mindset of adaptability further enhance this synergy, enabling creators to transcend limitations and unlock their full potential.

In a world of rapid change and growing complexity, these strategies provide a roadmap for navigating the creative journey with confidence and curiosity. By embracing innovation while staying true to their values, creators can shape a future where creativity is boundless and transformative.

A Path to Unbounded Creativity

The journey to unbounded creativity is filled with challenges, yet it is through these very obstacles that the creative spirit is tested, refined, and elevated. Societal norms, self-imposed limitations, and structural barriers may appear as roadblocks, but they also serve as opportunities for innovation and growth. Overcoming these restrictions requires the courage to challenge conventions, the adaptability to embrace change, and the persistence to persevere despite setbacks.

Challenging Norms: Expanding Creative Boundaries

Cultural and societal norms often dictate what is considered acceptable or valuable in artistic expression. While these frameworks provide structure, they can also stifle experimentation and constrain creativity. However, history shows that groundbreaking innovation often arises from questioning and redefining these very norms.

For example, movements such as Impressionism or street art exemplify how defying conventions can lead to new artistic paradigms. Artists like the Impressionists challenged traditional academic standards, while contemporary figures like Banksy push boundaries through socially charged street art. These trailblazers demonstrate that creativity flourishes when creators have the courage to question established norms, paving the way for more inclusive and dynamic artistic landscapes.

Addressing Internal Barriers: Unlocking Creative Potential

External challenges are significant, but internal barriers—self-doubt, fear of failure, and perfectionism—can be even more limiting. These internal struggles often inhibit creators from pursuing bold ideas or sharing their work. Overcoming these obstacles is both transformative and essential for realizing one's creative potential.

The process begins with self-awareness. Recognizing the root of self-doubt and reframing failure as a necessary part of growth empowers creators to move forward. Practical strategies like setting small, attainable goals, seeking constructive feedback, and approaching challenges with curiosity rather than judgment can help mitigate these internal limitations. By confronting and addressing these barriers, creators gain confidence, resilience, and renewed inspiration to explore their craft fearlessly.

Fostering Inclusivity: Broadening the Creative Vision

Creativity thrives in an environment that embraces diversity and inclusivity. Incorporating underrepresented voices and diverse perspectives enhances artistic expression, challenges entrenched biases, and fosters deeper connections with audiences. Inclusivity not only enriches the creative process but also ensures that art reflects the full spectrum of human experience.

For instance, films featuring actors from diverse backgrounds or literature highlighting marginalized voices bring new dimensions to storytelling. Such efforts not only expand the cultural relevance of creative works but also foster empathy and understanding. By championing inclusivity, creators contribute to a richer, more authentic artistic ecosystem that resonates universally.

Navigating and Transcending Restrictions

The journey to unbounded creativity is not about eliminating all restrictions but about transforming them into opportunities for growth and innovation. Constraints, whether external or internal, often ignite resourcefulness and ingenuity in creators.

For example, filmmakers working within tight budgets frequently develop innovative storytelling techniques, while poets crafting within the constraints of a sonnet or haiku often produce deeply impactful works through disciplined creativity. These examples demonstrate that limitations can serve as fertile ground for invention when approached with an open and adaptive mindset.

Turning Challenges into Catalysts for Growth

Challenges, whether societal, structural, or personal, are not merely obstacles but potential stepping stones for greater innovation. Creators who embrace difficulties as integral to their journey often find that these experiences enrich their work and deepen their connection to their craft.

By confronting challenges head-on, creators develop resilience, adaptability, and a broadened perspective. These qualities not only enhance their ability to navigate future obstacles but also inspire others to overcome their own limitations, creating a ripple effect of creativity and growth.

Building a Future of Unbounded Creativity

The path to unbounded creativity lies in balance—honoring tradition while embracing innovation, addressing personal fears while fostering collective inclusion, and navigating existing norms while envisioning new possibilities. It is a journey that calls for intentionality, authenticity, and openness to change.

As creators challenge conventions, overcome internal barriers, and champion inclusivity, they contribute to a vibrant and diverse creative ecosystem. This collective effort not only unlocks individual potential but also inspires communities, pushing the boundaries of what creativity can achieve.

Ultimately, the journey to unbounded creativity is about recognizing challenges as opportunities for growth and transformation. By learning to navigate and transcend them, creators craft works that are as impactful and meaningful as the journey itself. Creativity, when freed from constraints, becomes a powerful force for connection, innovation, and positive change.

Summary

This chapter delves into the complex challenges faced by creators in the modern age, where technological advancements offer unprecedented opportunities while simultaneously raising critical questions about authenticity, ethics, and purpose. It explores how creators navigate these tensions and redefine their roles in a rapidly evolving landscape.

Key themes include:

Balancing Tradition and Innovation: Creators grapple with integrating cutting-edge tools like AI while preserving the authenticity and emotional depth of their work.

The chapter highlights the struggle to respect traditional methods without resisting technological progress.

Authenticity vs. Efficiency: It examines the tension between maintaining originality and leveraging technology for efficiency and scalability.

Examples show how creators often fear losing their unique voice or having their work overshadowed by machine-generated outputs.

Ethical and Societal Implications: Ethical dilemmas are discussed, such as biases in AI, copyright concerns, and the potential for technology to homogenize creative outputs.

The chapter stresses the need for creators to take responsibility for the societal impact of their work, ensuring inclusivity and fairness.

The Pressure to Adapt: Modern creators face an intense demand to stay relevant, constantly upskilling and experimenting with new tools.

This pressure is juxtaposed with the challenge of maintaining creative integrity and avoiding burnout in a hyper-competitive environment.

Redefining Success: The chapter questions conventional metrics of success in the digital age, urging creators to focus on meaning, impact, and connection over mere productivity or popularity.

It emphasizes that success should be aligned with personal values and a broader purpose.

Navigating the Creative Identity Crisis: Many creators experience a crisis of identity as they adapt to hybrid workflows involving both human and machine collaboration.

The chapter explores strategies to reconcile this duality, empowering creators to embrace technology as an ally rather than a threat.

This chapter provides a nuanced exploration of the dilemmas faced by modern creators, offering insights into how they can overcome these challenges while preserving the authenticity and purpose of their craft. It invites readers to reflect on their own creative journeys and redefine their approach in the context of evolving technological landscapes.

Part III: Restrictions and Their Consequences

"Restrictions, whether self-imposed or societal, are the silent architects of creativity—they can either confine the mind or challenge it to soar beyond boundaries."

— Sandeep Chavan

7: The Fear of Machines

The rise of artificial intelligence has sparked both awe and apprehension, particularly among organizations tasked with navigating its implications. While AI promises unprecedented efficiency, accuracy, and innovation, its adoption has been met with resistance across industries and institutions. This hesitation often stems from a fear of the unknown—concerns about ethical dilemmas, potential job displacement, and the perceived loss of human control. Organizations, bound by tradition and cautious policies, frequently find themselves torn between the promise of progress and the risks of disruption. Understanding this fear and its underlying causes is essential to fostering a balanced approach to AI adoption, one that aligns ethical considerations with the practical realities of an increasingly automated world.

7.1 Critique of Institutional AI Policies

The integration of artificial intelligence into organizational frameworks has been fraught with resistance, often rooted in historical patterns of skepticism toward new technologies. Institutional AI policies frequently reflect this hesitancy, resulting in overregulation that stifles innovation and limits AI's transformative potential. While caution is necessary to address ethical and societal concerns, overly restrictive policies can hinder progress, leaving organizations unprepared for the future.

Historical Reluctance: Institutions and Technological Change

Institutions have long exhibited a tendency to resist disruptive technologies, often prioritizing stability and continuity over the uncertainties of innovation. This cautious approach aims to preserve established structures but frequently delays the adoption of advancements that ultimately prove transformative. The rise of artificial intelligence (AI) is a modern example of this historical pattern, echoing past apprehensions about mechanization, automation, and their societal implications.

The Fear of Disruption

Technological innovation has historically been met with skepticism and resistance, driven by fears of economic and cultural upheaval. During the Industrial Revolution, mechanized systems sparked anxiety among factory owners over transition costs and among workers who feared losing their livelihoods. Movements like the Luddites, who destroyed weaving machines in protest, exemplify how disruptive technologies can ignite resistance rooted in uncertainty and perceived threats to stability.

Similar apprehensions accompanied the advent of computers in the mid-20th century. Critics argued that computerization would dehumanize workplaces and lead to widespread job displacement. Early adoption was met with hesitation, as organizations grappled with the cultural and structural shifts required to integrate these technologies. Over time, however, computers revolutionized industries, demonstrating their indispensable value.

Parallels with AI Adoption

AI has inherited many of the fears associated with earlier disruptive technologies. Concerns about job displacement, loss of human oversight, and ethical dilemmas mirror historical anxieties about mechanization and computerization. Institutions often view AI as a double-edged sword—offering unprecedented efficiency and innovation while challenging established practices and power dynamics.

For example, organizations may hesitate to implement AI-driven automation, fearing employee backlash or stakeholder resistance over workforce reductions. Similarly, professions rooted in human expertise, such as law and medicine, wrestle with delegating critical decision-making tasks to algorithms. This reluctance underscores the tension between embracing progress and protecting traditional roles and systems.

The Cost of Delayed Adoption

While caution allows institutions to mitigate risks, excessive hesitation often comes at a significant cost. Delaying the adoption of transformative technologies like AI can result in missed opportunities, reduced competitiveness, and slower progress. Industries that were slow to embrace computerization, for instance, lost market share and efficiency to more agile competitors.

For AI, delayed adoption could hinder organizations from capitalizing on advancements that streamline operations, enhance decision-making, and improve customer experiences. Early adoption facilitates learning and adaptation, enabling organizations to address challenges, refine applications, and stay ahead in rapidly evolving markets. Prolonged resistance risks leaving institutions lagging in an increasingly AI-driven world.

The Need for Balance

Historical reluctance to embrace technological change underscores the importance of balance in institutional decision-making. While it is essential to consider the risks and societal implications of new technologies, it is equally critical to recognize their potential to drive progress and solve complex problems. Institutions that adopt a proactive yet measured approach are better positioned to navigate transformative eras like the rise of AI.

Phased implementation strategies, for example, allow organizations to address workforce concerns, ethical considerations, and technical challenges while gradually integrating AI into their operations. This mirrors the adoption of computers, where initial hesitations gave way to widespread acceptance as their benefits became undeniable. A balanced approach ensures that caution does not stifle innovation.

Learning from History

The patterns of resistance observed in previous technological revolutions provide valuable lessons for navigating AI's rise. They highlight the importance of education, collaboration, and open dialogue to address fears and misconceptions. Institutions that foster a culture of informed decision-making can mitigate risks while unlocking the opportunities presented by AI.

For instance, engaging stakeholders through workshops and training programs can build confidence and understanding of AI's potential. Transparent communication about the benefits and limitations of the technology can help dispel misconceptions and promote trust. By learning from historical precedents, institutions can strike a balance between caution and curiosity, ensuring progress without unnecessary delays.

Adapting to Thrive

History demonstrates that resistance to technological change is natural, but progress is inevitable. Institutions that adapt and evolve by addressing legitimate concerns while embracing innovation will not only survive but thrive. The integration of AI presents an opportunity to shape the future of innovation and creativity, provided institutions approach it with openness and strategic intent.

By navigating these challenges thoughtfully, institutions can position themselves as leaders in the transformative age of AI—leveraging its potential to enhance efficiency, improve decision-making, and create new opportunities. In doing so, they uphold a tradition of progress that has defined every technological revolution, ensuring that they remain relevant and resilient in the face of change.

Overregulation and Innovation Stagnation

In an effort to address ethical concerns and mitigate risks, institutions often prioritize caution over exploration, adopting policies that emphasize control and compliance. While these measures aim to ensure safety, fairness, and accountability, they can inadvertently stifle innovation by imposing excessive restrictions. Overregulation, particularly in emerging fields like artificial intelligence (AI), creates barriers to experimentation and implementation, limiting the transformative potential of these technologies.

The Cost of Excessive Caution

Overregulation frequently arises from the desire to prevent misuse or unintended consequences, especially in areas where ethical considerations intersect with technological capabilities. However, overly restrictive policies can have the unintended consequence of discouraging creativity and delaying the adoption of beneficial innovations. Institutions focusing solely on minimizing risks may forgo opportunities to harness AI's full potential.

For example, regulatory bodies often impose lengthy approval processes for AI systems in critical sectors such as healthcare or finance. While these safeguards aim to ensure accuracy and prevent harm, they can delay the deployment of technologies capable of saving lives or improving operational efficiency. This bottleneck discourages innovators, slows progress, and limits the societal benefits of AI.

Data Privacy Laws: Balancing Protection and Progress

Stringent data privacy regulations, like the General Data Protection Regulation (GDPR) in the European Union, are essential for protecting individual rights in the digital age. However, these laws also pose significant challenges for AI development, which depends on large volumes of high-quality data to function effectively. Privacy restrictions often limit access to such datasets, impeding AI's potential to drive progress.

For instance, healthcare institutions face barriers to sharing patient data for research, even when using anonymization techniques. These restrictions hinder the development of AI-driven diagnostics, personalized treatments, and drug discovery. Similarly, businesses aiming to use customer data for service improvements may struggle with compliance burdens, slowing innovation and reducing competitiveness. The challenge lies in protecting privacy without compromising progress.

The Innovation Paradox

Regulation plays a dual role in fostering trust and accountability while potentially hindering innovation. On one hand, robust oversight ensures ethical practices and builds public confidence in AI technologies. On the other, rigid policies can discourage experimentation, delay adoption, and stifle creativity.

For example, financial institutions face strict compliance requirements for AI algorithms used in fraud detection or credit risk assessment. While these regulations aim to ensure fairness and reliability, they often slow the agility needed to respond to emerging opportunities. The paradox highlights the need to strike a balance between regulatory oversight and the flexibility required for innovation.

The Real-World Impact of Overregulation

Overregulation has tangible consequences in industries where AI's potential remains underutilized due to restrictive policies.

✓ In the pharmaceutical sector, the prolonged approval process for AI-driven drug discovery models delays the development of critical treatments.

✓ Autonomous vehicles, despite their potential to reduce accidents and improve transportation efficiency, face significant regulatory hurdles that slow their integration into public systems.

Conversely, sectors with adaptive regulatory approaches have witnessed substantial advancements. For instance, regulatory sandboxes in fintech allow companies to test AI solutions in controlled environments, enabling rapid innovation while maintaining oversight. These examples illustrate the importance of balancing caution with flexibility to maximize progress.

Striking a Balance for Progress

To address the challenges of overregulation, institutions must design policies that balance caution with exploration. Adaptive frameworks, such as phased implementation and iterative evaluations, enable organizations to test AI technologies while addressing ethical concerns and minimizing risks. Collaboration between policymakers, industry leaders, and researchers is crucial to crafting regulations that are both robust and flexible.

Transparency and education are also essential in fostering trust and reducing resistance to AI adoption. By promoting a clear understanding of AI's capabilities and limitations, institutions can create an environment where innovation thrives alongside accountability. Engaging stakeholders in open dialogue ensures that regulatory frameworks remain aligned with technological advancements and societal priorities.

Unlocking AI's Potential Responsibly

While overregulation can hinder progress, thoughtful oversight remains essential for ensuring ethical AI deployment. The goal should not be to eliminate regulation but to craft policies that enable innovation while safeguarding societal values. Adaptive and collaborative approaches can unlock AI's transformative potential, allowing its benefits to be realized without compromising safety or integrity.

By addressing risks without stifling creativity, institutions can foster an ecosystem where progress flourishes. In doing so, they position themselves to lead in shaping a future where technology is leveraged responsibly and effectively for humanity's best interests. Striking the right balance will ensure that AI becomes a tool for innovation and growth, rather than a casualty of excessive caution.

Balancing Progress with Caution: Navigating the AI Policy Dilemma

Striking a balance between ethical responsibility and technological progress is one of the most pressing challenges in the era of artificial intelligence (AI). While addressing concerns like bias, privacy, and accountability is crucial, excessive caution can stifle innovation and prevent society from benefiting fully from AI's transformative potential. Achieving this equilibrium requires policies that carefully weigh risks against rewards, fostering advancements while safeguarding societal values.

The Dual Mandate of Institutions

Institutions face a dual mandate: to protect public interests and to promote innovation. Unchecked AI development risks ethical violations, data misuse, and unforeseen societal harm. Conversely, overly restrictive policies can slow progress, reduce competitiveness, and limit the benefits AI offers.

For instance, AI in healthcare holds immense promise for diagnostics and personalized medicine, yet lengthy regulatory approval processes delay its application. Similarly, in education, adaptive learning systems could revolutionize personalized instruction, but concerns over data privacy and compliance often impede their widespread adoption. These examples underscore the delicate balancing act institutions must navigate to advance innovation without compromising ethical standards.

The Promise of Adaptive Policies

Adaptive policies provide a dynamic framework to address the rapid evolution of AI technologies. Unlike static regulations, they evolve alongside advancements, allowing institutions to respond to new challenges while maintaining oversight.

One effective tool is the use of **regulatory sandboxes**—controlled environments where organizations can test AI applications with reduced regulatory hurdles. These sandboxes enable innovators to identify and address potential risks while refining their technologies before full-scale deployment. This iterative approach ensures both innovation and compliance, reducing bureaucratic delays while safeguarding public interest.

Fostering Responsible Experimentation

Responsible experimentation is key to balancing progress with caution. Institutions can encourage innovation by creating environments that prioritize open dialogue, collaboration, and iterative development. Partnerships between policymakers, industry leaders, and researchers can lead to regulations that are robust yet flexible, embedding ethical considerations into AI development from the outset.

Transparency is central to responsible experimentation. Clear communication about the objectives, methods, and potential risks of AI projects builds trust with stakeholders and the public. Transparency also invites diverse feedback, reducing the likelihood of unintended consequences and ensuring that AI applications align with societal needs and values.

Case Studies: Balanced Approaches in Action

Certain industries offer models for achieving a balance between progress and caution.

Financial Sector: AI-driven fraud detection systems operate under regulatory frameworks that emphasize accuracy and accountability. Regular audits and compliance checks ensure ethical application while enabling technological advancements.

Autonomous Vehicles: Phased testing and deployment strategies have allowed developers to address safety concerns incrementally. Controlled trials provide valuable data for refining algorithms, gradually building public confidence and regulatory support.

These examples highlight the potential of measured approaches that prioritize both innovation and responsibility.

Challenges and Opportunities

Balancing progress and caution is not without challenges. Policymakers must reconcile competing priorities, such as safeguarding individual rights while driving economic growth. Additionally, global inconsistencies in regulatory approaches can lead to uneven progress, with some regions advancing rapidly while others lag due to excessive restrictions.

However, these challenges also present opportunities for collaboration and innovation. International cooperation on AI governance can establish shared ethical standards and best practices, ensuring that technological progress benefits humanity collectively. By learning from successful initiatives and iterating on policies, institutions can refine their strategies and foster sustainable AI ecosystems.

A Vision for Balanced Progress

Achieving balanced progress is not about eliminating risks but managing them intelligently. Adaptive policies, transparent experimentation, and collaborative networks provide a roadmap for navigating the complexities of AI integration while maintaining ethical integrity.

This approach ensures AI becomes a force for good, driving innovation and improving lives without compromising societal values. By adopting this balanced path, institutions can shape a future where technological advancements and ethical responsibilities coexist, unlocking AI's full potential for humanity's benefit.

Real-World Impacts: Case Studies of AI Policy Resistance

Institutional resistance to artificial intelligence (AI) often results in delayed adoption, missed opportunities, and inefficiencies. While caution is typically motivated by the desire to uphold ethical and societal values, excessive hesitation can have significant consequences, particularly in sectors where timely innovation has the potential to save lives, lower costs, or improve services. Real-world case studies provide valuable insights into the challenges posed by resistance and the rewards of adopting balanced policies that integrate ethical oversight with progress.

Healthcare: The Cost of Delayed AI-Driven Diagnostics

AI-powered diagnostic tools have demonstrated the ability to analyze medical data with remarkable speed and accuracy, enabling early detection and personalized treatment of diseases. For example, AI algorithms can identify subtle patterns in imaging scans, offering earlier diagnoses for conditions like cancer and heart disease. However, regulatory hurdles, such as lengthy approval processes and stringent privacy requirements, have slowed the adoption of these tools.

During the COVID-19 pandemic, AI models capable of predicting virus spread, optimizing resource allocation, and assisting in vaccine distribution faced similar regulatory delays. These bottlenecks hindered the rapid deployment of potentially life-saving solutions, underscoring how overcautious policies can impede timely responses during public health crises. Such delays highlight the need for frameworks that expedite innovation without compromising safety.

Public Services: Stalled Automation and Missed Efficiencies

AI-driven automation in public services, such as waste management, traffic optimization, and administrative processes, holds the promise of enhancing efficiency and improving citizen experiences. Yet, resistance from institutions often leads to delayed implementation, perpetuating inefficiencies and increasing costs.

For instance, smart city initiatives aimed at using AI to optimize urban infrastructure frequently encounter pushback due to concerns over data surveillance and potential misuse. While these concerns are valid, excessive resistance has slowed projects that could reduce traffic congestion, lower energy consumption, and enhance public safety. These missed opportunities highlight the importance of balancing ethical safeguards with the urgency of innovation.

Education: Hesitation in Adopting Personalized Learning Tools

In education, AI-powered adaptive learning platforms have shown great promise in tailoring instruction to individual student needs, improving engagement and learning outcomes. Despite their potential, concerns about data privacy, security, and equity have hindered widespread adoption.

For example, some schools have resisted implementing AI tools out of fear that they might exacerbate educational inequalities or compromise student privacy. While these ethical considerations are important, the reluctance to adopt AI has prevented many students, particularly in underfunded schools, from accessing resources that could help bridge achievement gaps. The lack of integration leaves significant opportunities for educational transformation untapped.

Success Stories: Balanced Policies in the Financial Sector

Industries that have adopted balanced approaches to AI demonstrate how ethical oversight and technological innovation can coexist. The financial sector is a prime example, leveraging AI for fraud detection, credit risk assessment, and investment analysis under well-defined regulatory frameworks.

AI-powered fraud detection systems, for instance, analyze vast datasets to identify suspicious transactions in real time, significantly reducing financial crime. These systems operate within strict guidelines that prioritize accuracy, accountability, and fairness, showcasing how adaptive policies can enable innovation while safeguarding public trust. This success underscores the importance of fostering progress through thoughtful regulation.

Transportation: Pragmatic Integration of Autonomous Vehicles

The transportation sector offers another example of balancing caution with progress. Autonomous vehicles (AVs) have faced intense scrutiny due to safety and liability concerns. However, rather than halting development, many jurisdictions have adopted phased testing and deployment strategies.

For instance, AVs are being trialed in controlled environments such as university campuses and designated urban districts. These pilots allow developers to refine their technologies and address safety concerns while generating data to inform future policies. This measured approach demonstrates how institutions can manage risks without stifling innovation, paving the way for broader adoption in the future.

Key Lessons: Achieving Balance in AI Policymaking

The case studies reveal a consistent theme: resistance to AI often stems from valid ethical and societal concerns, but excessive caution can hinder innovation and limit progress. Conversely, industries that adopt balanced policies—addressing risks while enabling experimentation—are better positioned to unlock AI's transformative potential.

Collaboration among policymakers, industry leaders, and researchers is critical to achieving this balance. Transparent and adaptive regulatory frameworks, coupled with stakeholder engagement, ensure that AI deployment aligns with ethical standards while fostering progress. International cooperation can further harmonize approaches, enabling consistent and equitable advancements across regions.

A Way Forward: Embracing AI for Societal Good

Through thoughtful policymaking, institutions can move beyond resistance and leverage AI as a tool for innovation, efficiency, and societal benefit. By learning from the pitfalls of excessive caution and the successes of balanced approaches, organizations can shape a future where technology serves humanity's best interests.

Striking the right balance between caution and progress ensures that AI is deployed responsibly, unlocking its potential to drive transformative change while safeguarding societal values. This equilibrium is essential for fostering trust, enhancing innovation, and realizing the full promise of AI.

Institutional AI Policies: Shaping the Future of Technological Integration

Institutional policies on artificial intelligence (AI) play a crucial role in shaping how this transformative technology integrates into society. These policies influence not only the speed of innovation but also its ethical, societal, and practical impacts. While addressing concerns such as bias, privacy, and accountability is essential, excessive regulation or resistance to AI can hinder progress, stifling creativity and limiting its potential to solve complex problems and improve lives.

Learning from the Past

History provides valuable insights into the consequences of resisting disruptive technologies. During the industrial revolution, institutions that adopted adaptable policies successfully navigated the societal shifts brought on by mechanization, while rigid resistance led others to lag behind. Similarly, the digital age underscored the importance of embracing innovation alongside thoughtful regulation, with early adopters of computing technologies gaining significant advantages.

AI presents a comparable challenge. Striking the right balance between safeguarding societal values and fostering an environment that encourages experimentation and growth is critical. Institutions must avoid the pitfalls of rigidity, which can lead to missed opportunities, and instead adopt policies that allow for flexibility and adaptation as AI evolves.

Adaptive Policies: A Blueprint for Progress

Adaptive policies are central to successfully integrating AI into society. These frameworks evolve alongside technological advancements, addressing risks without stifling innovation.

One effective approach is the use of **regulatory sandboxes**—controlled environments where organizations can test AI applications under real-world conditions while remaining within regulatory oversight. By identifying potential risks and refining systems in a safe setting, sandboxes allow innovators to experiment responsibly. This method builds trust among stakeholders, fosters public confidence, and ensures that AI's benefits are accessible and equitable.

Additionally, adaptive policies can facilitate phased implementation strategies, enabling gradual deployment of AI systems. For instance, testing autonomous vehicles in limited, controlled environments before broader rollout has proven effective in addressing safety concerns while maintaining progress.

Fostering Responsible Innovation

A culture of responsible innovation is vital to the success of institutional AI policies. Transparency, inclusivity, and collaboration must be prioritized to ensure that AI development aligns with societal values and addresses diverse needs.

Transparency: Open communication about AI's capabilities, limitations, and potential risks fosters trust among stakeholders. Publicly sharing goals and methodologies allows institutions to address concerns proactively.

Inclusivity: Engaging a wide range of perspectives—across cultural, economic, and professional spectrums—ensures that AI systems are designed to serve diverse populations effectively. Inclusivity also helps identify and mitigate unintended biases in AI applications.

Collaboration: Partnerships between policymakers, industry leaders, and researchers create a robust ecosystem for AI development. Collaborative efforts facilitate the alignment of ethical standards with technological goals, ensuring that policies are both practical and forward-thinking.

Unlocking AI's Transformative Potential

Institutions that embrace adaptive policies and foster responsible innovation position themselves to unlock AI's transformative potential. When implemented thoughtfully, AI can address some of humanity's most pressing challenges—from improving healthcare diagnostics to optimizing energy consumption and enhancing education.

By balancing ethical oversight with an openness to experimentation, institutions can pave the way for AI systems that are not only innovative but also equitable and trustworthy. This approach ensures that the benefits of AI are shared broadly, enhancing quality of life while upholding societal values.

A Future of Equitable Progress

Institutional AI policies must evolve to meet the demands of a rapidly changing technological landscape. Striking the right balance between caution and progress is not merely a regulatory challenge but a moral imperative. Policies that are adaptive, inclusive, and collaborative enable institutions to navigate the complexities of AI integration while ensuring that its deployment aligns with humanity's best interests.

By learning from the past, embracing adaptive strategies, and fostering a culture of responsible innovation, institutions can shape a future where AI serves as a force for good—driving progress, enhancing equity, and improving lives worldwide.

7.2 Ethical Concerns vs. Practical Realities

The integration of artificial intelligence (AI) into society has sparked an ongoing debate between ethical concerns and practical realities. While AI offers transformative potential, it also raises significant questions about fairness, privacy, accountability, and control. Institutions face the dual challenge of addressing these ethical dilemmas while leveraging AI's capabilities to improve efficiency and innovation. Understanding and balancing these competing priorities is essential for navigating the complexities of AI adoption.

The Ethics Debate: Navigating Bias, Privacy, and Control

Ethical concerns surrounding artificial intelligence (AI) focus on critical issues like bias, privacy, and control, raising fundamental questions about its role in society. While AI holds immense promise, its reliance on human-created data and algorithms makes it susceptible to inherent flaws, demanding thoughtful examination and proactive solutions.

Bias in AI: Reinforcing Inequality

Despite its reputation as an objective tool, AI is only as unbiased as the data and algorithms that power it. When training datasets reflect societal prejudices or lack diversity, AI systems can unintentionally amplify discrimination rather than mitigate it.

For instance, hiring algorithms designed to streamline recruitment processes have shown biases against certain demographics, mirroring historical inequities embedded in training data. This perpetuates unfair hiring practices, undermining the promise of equity in automation.

Similarly, in criminal justice, predictive policing tools have disproportionately targeted marginalized communities, reinforcing existing disparities. Facial recognition technologies have also exhibited higher error rates for individuals with darker skin tones, calling into question their reliability and fairness.

To address these challenges, diverse and representative datasets must become a priority, alongside rigorous testing and validation to identify and mitigate bias before deployment. Such measures ensure that AI systems enhance fairness and inclusion rather than perpetuate inequality.

Privacy Concerns: Safeguarding Personal Data

AI's effectiveness often relies on vast amounts of personal data, raising significant privacy concerns. From medical records to social media activity, the data used by AI systems can expose individuals to breaches, misuse, or surveillance.

For example, advertising platforms leveraging AI to analyze user behavior may inadvertently reveal private habits or preferences, creating risks of exploitation. Similarly, healthcare applications reliant on patient data risk breaches that could compromise both privacy and trust in AI-powered solutions.

Robust privacy safeguards are essential to mitigate these risks. Techniques such as data anonymization, encryption, and differential privacy can protect individuals while enabling innovation. Clear policies, ethical data collection practices, and transparent communication about how data is used are equally critical for maintaining public trust in AI systems.

Control and Accountability: Navigating Autonomy

As AI systems gain autonomy, questions about control and accountability become increasingly complex. Determining responsibility for AI-driven decisions—whether in autonomous vehicles, loan approvals, or medical diagnoses—is a growing ethical challenge.

For instance, if an autonomous vehicle causes an accident, who should bear responsibility—the developer, the manufacturer, or the operator?

Similarly, ethical dilemmas, such as prioritizing lives in unavoidable harm scenarios, highlight the challenges of encoding human values into AI systems.

Human oversight remains vital in decision-making processes. While AI enhances efficiency and accuracy, overreliance without adequate human review can lead to significant ethical oversights. Ensuring that humans remain integral to critical decisions helps maintain accountability and trust in AI systems.

Establishing Ethical Frameworks

The ethical challenges of AI underscore the need for robust frameworks that prioritize transparency, inclusivity, and accountability. These frameworks ensure AI systems align with societal values while addressing risks and inequities.

✓ **Transparency**: Making AI algorithms and decision-making processes understandable allows users and stakeholders to trust and challenge AI outcomes.

✓ **Inclusivity**: Engaging diverse perspectives ensures fairness, reduces bias, and improves the equity of AI systems.

✓ **Accountability**: Establishing clear guidelines for responsibility when AI systems cause harm or violate ethical standards is essential.

Initiatives such as "explainable AI" provide insights into how decisions are made, empowering users to question and refine outcomes. Ethical review boards and independent audits can ensure responsible development and deployment, creating mechanisms to address and rectify issues proactively.

Balancing Innovation and Responsibility

Addressing ethical concerns like bias, privacy, and control is essential for harnessing AI's transformative potential. By integrating ethical considerations into every stage of AI development, institutions can build systems that are innovative, fair, and equitable.

A proactive, transparent approach fosters trust and ensures that AI serves humanity effectively, aligning technological progress with societal values. Striking this balance will be critical to unlocking AI's benefits while safeguarding its responsible integration into the fabric of our lives.

Practical Necessities: AI's Role in Efficiency and Accuracy

While ethical concerns surrounding artificial intelligence (AI) are valid and essential to address, the transformative practical benefits AI offers are equally significant. AI's unparalleled ability to process vast datasets, identify patterns, and make precise decisions enables it to solve problems at a scale and speed beyond human capability. Across industries, AI has become a cornerstone for driving efficiency, enhancing accuracy, and fostering innovation.

Healthcare: Revolutionizing Diagnostics and Treatment

AI is redefining healthcare by transforming how diseases are diagnosed, treated, and managed. AI-powered diagnostic tools analyze medical images, genetic data, and patient histories to detect conditions earlier and with greater accuracy than traditional methods. For instance, AI algorithms in radiology can identify subtle signs of cancer or fractures in imaging scans that might be missed by even the most experienced clinicians. Early detection leads to timely interventions, improving patient outcomes and saving lives.

Beyond diagnostics, AI accelerates drug discovery by identifying potential treatments and reducing research timelines from years to months. Administrative tasks, such as scheduling appointments and triaging patients, are streamlined through AI systems, allowing healthcare professionals to dedicate more time to patient care. These advancements illustrate how AI enhances not only efficiency but also the scope of medical innovation.

Logistics and Supply Chain Management: Optimizing Operations

AI has revolutionized logistics and supply chain management by optimizing operations, reducing costs, and improving sustainability. AI algorithms analyze extensive datasets to determine the most efficient transportation routes, minimizing delivery times and fuel consumption. For instance, AI-powered route optimization tools enable delivery companies to adapt in real-time to traffic conditions, ensuring timely deliveries while lowering environmental impact.

In inventory management, predictive analytics allow businesses to forecast demand with remarkable accuracy, avoiding overstocking and shortages. This capability is particularly critical in industries like e-commerce and retail, where customer satisfaction depends on reliable and efficient supply chains. By improving operational workflows, AI enhances economic performance while supporting sustainability goals.

Finance: Enhancing Security and Decision-Making

In the financial sector, AI drives innovation in security, decision-making, and operational efficiency. AI-powered fraud detection systems analyze transactions in real-time, identifying anomalies that signal potential fraud. These systems not only safeguard assets but also bolster customer trust by ensuring secure transactions.

AI's role in investment management is equally transformative. Algorithms evaluate market trends, economic indicators, and historical data to provide actionable insights for crafting investment strategies. Robo-advisors democratize financial planning by offering personalized services to individuals, making wealth management more accessible. By combining efficiency with precision, AI helps financial institutions navigate complex markets and deliver superior outcomes for clients.

AI as a Catalyst for Broader Innovation

AI's influence extends far beyond individual industries, driving progress in fields such as renewable energy, education, and scientific research.

✓ **Energy**: AI optimizes energy grids by predicting demand and managing supply, ensuring efficient resource utilization.

✓ **Education**: AI-powered learning platforms adapt to individual students' needs, enhancing engagement and academic outcomes.

✓ **Scientific Research**: AI accelerates discovery, from modeling climate change scenarios to designing sustainable materials, addressing some of the world's most pressing challenges.

These applications highlight AI's role as a necessity, not merely a convenience, for tackling global problems and fostering innovation.

Balancing Practical Benefits with Ethical Standards

The challenge lies in harnessing AI's practical benefits while addressing ethical concerns such as data privacy, bias, and accountability. Achieving this balance requires transparent policies, robust governance frameworks, and inclusive datasets to ensure that AI serves society responsibly.

✓ **Healthcare**: Anonymizing patient data can alleviate privacy concerns while preserving AI's ability to deliver life-saving insights.

✓ **Finance**: Regular audits of AI algorithms ensure that decision-making processes remain fair and unbiased.

✓ **Education**: Safeguarding student data and ensuring equitable access to AI-driven tools promote ethical and inclusive implementation.

By prioritizing both efficiency and ethical integrity, institutions can leverage AI to drive transformative progress while addressing societal challenges.

AI's Indispensable Role in the Modern World

AI's ability to enhance accuracy, streamline operations, and accelerate innovation makes it a vital tool for addressing the complexities of the modern world. The key to realizing its full potential lies in embracing its capabilities while remaining vigilant about its challenges.

Through thoughtful integration and proactive governance, AI can be a force for good, enabling breakthroughs that improve lives, advance industries, and foster a more equitable future. Balancing progress with responsibility ensures that AI serves humanity effectively and inclusively.

Misaligned Priorities: When Fear Overrides Opportunity

A major challenge in adopting artificial intelligence (AI) lies in misaligned priorities, where fear of potential risks overshadows recognition of its transformative potential. Ethical concerns—such as bias, privacy, and accountability—are undeniably important, but when these issues dominate the narrative, they often lead to excessive caution. This overly cautious approach can stifle innovation, delay progress, and hinder the development of solutions urgently needed to address global challenges.

The Cost of Overregulation

One of the clearest examples of misaligned priorities is the imposition of stringent regulations aimed at preventing data misuse. While these policies are designed to protect privacy and ensure ethical AI use, they often limit access to the large and diverse datasets critical for training and refining AI systems.

For instance, in healthcare, restrictive data-sharing policies can delay advancements in AI-powered diagnostics. These tools have the potential to detect diseases earlier, optimize treatments, and improve public health outcomes. Similarly, in climate science, strict data regulations hinder the development of AI models that predict weather patterns, assess environmental risks, or optimize renewable energy systems. The delays caused by such fear-driven policies can leave society underprepared to address urgent challenges, highlighting the unintended consequences of prioritizing caution over progress.

Automation Anxiety and Job Displacement Fears

Fears surrounding job displacement due to automation provide another example of imbalanced priorities. Concerns about AI replacing human labor often dominate public discourse, fueling resistance to its adoption across industries like manufacturing, retail, and logistics. While the potential for disruption exists, these fears frequently overshadow discussions about how AI can complement human labor, enhance productivity, and create new opportunities.

AI-driven automation can free workers from repetitive and labor-intensive tasks, enabling them to focus on roles that require creativity, problem-solving, and interpersonal skills. For instance, in agriculture, AI-enabled equipment can improve efficiency and reduce the physical burden on farmers while increasing crop yields. Addressing automation anxiety requires reframing the narrative to emphasize how AI can augment human capabilities, drive economic growth, and support reskilling initiatives to prepare workers for the jobs of tomorrow.

The Ripple Effect of Imbalanced Policies

When fear dominates policy-making, it creates a ripple effect that stifles innovation across entire sectors. Institutions implementing overly restrictive policies may inadvertently discourage experimentation, limit collaboration, and reduce funding for AI research and development. This cautious mindset can result in missed opportunities to improve public services, drive technological progress, and enhance quality of life.

For example, concerns about privacy and bias have slowed the adoption of AI in education. Adaptive learning platforms, designed to tailor instruction to individual students' needs, have shown significant promise in improving engagement and outcomes. However, resistance to their widespread implementation deprives students of personalized educational experiences that could bridge achievement gaps and foster equitable learning. Such cases demonstrate the importance of finding a balanced approach that enables innovation while addressing legitimate concerns.

Striking the Right Balance

Overcoming misaligned priorities requires a nuanced approach that weighs both the risks and rewards of AI adoption. Striking this balance involves fostering a culture of responsible innovation, where ethical considerations are addressed without stifling creativity.

Adaptive regulatory frameworks, such as regulatory sandboxes, provide an effective solution. These controlled environments allow organizations to test AI applications, identify risks, and refine their systems while maintaining the freedom to innovate. Such frameworks promote experimentation while ensuring accountability.

Collaboration is another critical factor. Policymakers, industry leaders, and researchers must engage in open dialogue to create policies informed by diverse perspectives and aligned with societal values. Transparent communication about AI's benefits and limitations can help build public trust, reducing resistance and encouraging broader acceptance of new technologies.

Moving Forward with Purpose

The path forward lies in aligning priorities to balance ethical concerns with the practical realities of AI integration. By acknowledging and addressing risks without allowing them to overshadow opportunities, institutions can unlock AI's transformative potential while maintaining ethical integrity.

This balanced approach ensures that AI becomes a tool for societal progress, empowering humanity to address complex challenges with confidence and purpose. Through adaptive policies, transparent collaboration, and a commitment to responsible innovation, institutions can pave the way for a future where AI drives meaningful and inclusive advancements.

Bridging the Divide: Ethical AI in Action

The perceived divide between ethical concerns and practical necessities in artificial intelligence (AI) adoption is far from insurmountable. Ethical AI offers a way to reconcile these priorities, addressing critical issues like bias, privacy, and accountability while enabling AI to deliver meaningful societal benefits. Rather than a compromise, ethical AI serves as a pathway to balance, fostering innovation without sacrificing foundational principles.

Designing Transparent, Fair, and Accountable Systems

Transparency is a cornerstone of ethical AI. By making algorithms understandable and accessible, institutions can build trust among users, regulators, and stakeholders. **Explainable AI (XAI)** exemplifies this approach, shedding light on how decisions are made and enabling the identification of errors or biases. This clarity empowers users to trust AI systems while ensuring ongoing refinement and accountability.

Fairness in AI systems hinges on diverse and representative datasets. Designing algorithms that consider varied demographics helps minimize bias and ensures equitable outcomes. Additionally, accountability mechanisms such as regular audits and clearly defined responsibilities maintain the integrity of AI systems, ensuring that developers and deployers are held responsible for the decisions and outcomes their systems generate.

Innovative Approaches to Data Privacy and Utility

One of AI's most significant ethical challenges is balancing data privacy with the need for robust datasets. Techniques like **data anonymization** and **differential privacy** safeguard individual identities while enabling valuable data to train AI systems effectively. These approaches bridge the gap between protecting personal information and driving technological progress.

For instance, healthcare organizations can anonymize patient data while using it to train AI models for disease diagnostics and treatment optimization. This dual focus ensures that ethical considerations do not obstruct advancements capable of saving lives and improving healthcare outcomes.

Collaborative Governance Models

Bridging the ethical-practical divide requires collaboration among diverse stakeholders, including policymakers, technologists, community representatives, and ethicists. Collaborative governance models provide platforms for inclusive dialogue, ensuring that AI development and deployment reflect societal values and address community needs.

AI ethics committees, comprising members from various sectors, can oversee projects to ensure compliance with ethical standards while fostering innovation. Public consultations and feedback mechanisms enhance inclusivity, giving individuals and communities a voice in shaping AI policies and applications. This collaborative approach fosters trust, accountability, and shared ownership of AI's impact on society.

Real-World Applications of Ethical AI

Ethical AI's transformative potential is evident in real-world applications that balance responsibility with innovation:

- ✓ **Disaster Response Systems:** AI-driven models analyze vast datasets to predict and mitigate the effects of natural disasters. Guided by ethical principles, these systems prioritize human safety and minimize harm.

- ✓ **Agriculture:** AI optimizes resource use while protecting the environment, using transparent algorithms to balance productivity with sustainability.

✓ **Education:** Adaptive learning platforms leverage anonymized data to personalize instruction for students, ensuring privacy while improving engagement and outcomes.

These examples demonstrate that ethical integrity and practical utility can coexist, creating solutions that are both impactful and responsible.

A Balanced Approach to AI Integration

Ethical concerns and practical realities are two sides of the same coin. Addressing these challenges requires a balanced approach that upholds ethical principles while embracing technological progress.

Open forums, **public-private partnerships**, and transparent communication channels can facilitate discussions about AI's implications, fostering mutual understanding and trust. Tools like **regulatory sandboxes** allow organizations to test AI applications in controlled environments, identifying and resolving risks without hindering innovation.

Shaping a Meaningful and Equitable Future

Embracing ethical AI enables institutions to harness its transformative power responsibly, ensuring its benefits are distributed equitably. Ethical AI is not about eliminating risks entirely but about managing them intelligently, designing systems that uphold human values while addressing societal challenges.

As AI evolves, the task is not merely to keep pace with its advancements but to guide its development in alignment with societal goals. Through thoughtful collaboration, transparency, and innovative frameworks, institutions can shape a future where AI enhances lives, solves global challenges, and embodies the best of human ingenuity and ethics.

Summary

This chapter examines the underlying reasons behind organizational resistance to adopting AI and other advanced technologies, despite their potential to transform industries and drive innovation. It sheds light on the cultural, practical, and psychological barriers that organizations face in embracing machines and explores strategies to overcome them.

Key themes include:

Cultural Resistance: Organizations with deep-rooted traditions often perceive advanced technologies as threats to established workflows and practices.

Fear of disrupting legacy systems and losing the human touch in operations contributes to a reluctance to adopt machines.

Job Security Concerns: The fear of widespread job displacement due to automation remains one of the primary reasons for resistance.

The chapter discusses how employees and leaders alike worry about their relevance in an AI-driven landscape, fostering hesitancy in embracing change.

Ethical and Societal Accountability: Organizations often struggle with the ethical implications of deploying machines, such as biases in AI algorithms or potential misuse of technology.

Concerns about public perception and accountability further contribute to caution in implementation.

Lack of Awareness and Understanding: Many organizations lack a clear understanding of how AI can be effectively integrated into their operations.

This knowledge gap leads to skepticism about the benefits of machines and uncertainty about their long-term implications.

Financial and Resource Constraints: Implementing advanced technologies often requires significant investment in infrastructure, training, and upskilling.

Resource limitations and fear of not achieving a return on investment deter many organizations from pursuing machine integration.

The Myth of Machine Infallibility: Some organizations resist machines due to unrealistic expectations or misconceptions about their capabilities.

The chapter highlights instances where technology failed to deliver expected outcomes, reinforcing fears about its reliability.

Strategies for Overcoming Resistance: The chapter concludes with actionable strategies for addressing these fears, such as fostering a culture of adaptability, promoting AI literacy, and emphasizing the collaborative potential between humans and machines.

It calls for leadership to take an active role in reshaping mindsets, showcasing successful use cases, and aligning machine integration with organizational goals.

This chapter provides a comprehensive analysis of the fears and challenges organizations face in adopting machines, while offering insights into how these barriers can be dismantled. It emphasizes the importance of a balanced, informed approach to ensure that technology serves as an enabler of progress rather than a source of apprehension.

8. Universities and Old Lenses

Universities have historically played a pivotal role in shaping innovation, serving as hubs of research, education, and thought leadership. However, as artificial intelligence (AI) continues to revolutionize industries and societies, many academic institutions appear hesitant to fully embrace its transformative potential. Rooted in traditional frameworks and cautious policymaking, universities risk lagging behind the pace of technological advancement, inadvertently fostering societal divides. This chapter explores the loopholes in current AI policies, the impact of resistance, and actionable suggestions for proactive adaptation.

8.1 Analyzing Loopholes in AI Policies

Universities, as vital hubs of knowledge and innovation, play a pivotal role in shaping the future of artificial intelligence (AI). Their policies, which often emphasize ethical safeguards and risk mitigation, reflect a necessary commitment to addressing societal and individual impacts. However, these well-meaning frameworks can inadvertently become overly rigid, hindering progress and limiting universities' ability to harness AI's transformative potential. Examining the gaps in current AI policies reveals opportunities for improvement that could better align ethical responsibility with innovation.

Overregulation of Data Usage

Data is the foundation of AI development, yet stringent regulations around data usage often impede progress. Universities, in their effort to protect privacy and prevent misuse, frequently implement policies that restrict access to datasets essential for training and refining AI models.

For example:

✓ **Healthcare Research**: Ethical concerns about patient privacy can prevent access to anonymized medical records, delaying the development of life-saving AI-driven diagnostic tools.

✓ **Environmental Studies**: AI models aimed at predicting climate change impacts are often constrained by policies limiting the use of global datasets, slowing critical advancements in sustainability and disaster response.

These restrictions, though rooted in valid ethical considerations, underscore the need for adaptive policies. Solutions such as data anonymization, secure data-sharing protocols, and collaborations with trusted data custodians can strike a balance between privacy protection and research progress.

Outdated Curricula

One of the most glaring gaps in university AI policies is the failure to modernize curricula in line with the rapid evolution of the field. Many programs remain focused on theoretical foundations, offering limited opportunities for students to engage with the practical applications of AI.

For instance:

✓ **Practical Experience**: While students may learn about the theoretical underpinnings of machine learning algorithms, they often lack exposure to platforms like TensorFlow, PyTorch, or real-world datasets.

✓ **Interdisciplinary Approaches**: Few programs integrate AI with other domains, such as ethics, policy, or industry-specific knowledge, which are essential for understanding AI's broader implications.

To bridge this gap, universities should revamp their curricula to include project-based learning, industry-relevant case studies, and training in cutting-edge tools. Collaborations with technology firms and open-source initiatives can provide students with hands-on experience, better preparing them for the demands of the modern workforce.

Limited Industry Collaboration

Strong partnerships between academia and industry are crucial for driving AI innovation, yet many universities face bureaucratic and institutional barriers that limit such collaborations.

Common obstacles include:

✓ **Restrictive Intellectual Property Policies**: Industry leaders may hesitate to partner with universities due to concerns over ownership rights to jointly developed technologies.

✓ **Bureaucratic Approval Processes**: Lengthy and complex procedures discourage timely collaboration on cutting-edge projects.

✓ **Institutional Resistance**: Faculty members seeking industry partnerships often encounter a lack of administrative support or unclear pathways for engagement.

To address these challenges, universities can adopt more flexible collaboration models, such as:

✓ Establishing clear and equitable frameworks for intellectual property sharing.

✓ Streamlining administrative processes to expedite partnership approvals.

✓ Creating liaison offices dedicated to fostering academia-industry collaboration.

These measures not only enhance research and innovation but also expose students and faculty to real-world applications, providing invaluable insights into industry trends and practices.

The Reluctance to Adapt

The gaps in AI policies reflect a broader hesitation among universities to adapt to the rapid pace of technological advancement. While their commitment to ethics and responsibility is laudable, a cautious approach often results in missed opportunities to lead innovation and influence AI development in alignment with societal values.

To remain at the forefront of AI progress, universities must overcome this reluctance by:

✓ **Encouraging Experimentation**: Creating regulatory sandboxes or pilot programs for testing new ideas in a controlled environment.

✓ **Modernizing Institutional Frameworks**: Ensuring policies are flexible enough to evolve alongside AI advancements.

✓ **Promoting Cross-Disciplinary Engagement**: Facilitating collaborations across fields to address AI's multifaceted challenges.

Reclaiming Leadership in AI Innovation

By addressing these loopholes, universities can reaffirm their position as pioneers in the AI space, fostering an environment where ethical considerations and innovation coexist harmoniously. The challenge lies in achieving a balance—protecting individuals and communities while enabling the research, collaboration, and education needed to unlock AI's full potential.

Adapting to AI's rapid evolution is not merely an institutional responsibility; it is an opportunity for universities to lead by example, shaping the future of AI in ways that prioritize both progress and societal well-being.

8.2 How Resistance Fosters Societal Divides

Universities, as pillars of knowledge and innovation, have a profound responsibility to guide society through technological advancements. Yet, hesitation to fully embrace artificial intelligence (AI) creates significant societal divides, perpetuating inequalities, and hindering progress in critical areas. Resistance to AI adoption not only delays innovation but actively widens existing gaps in access, opportunity, and global competitiveness.

Technological Inequity: Ill-Prepared Graduates

By resisting the integration of AI into their curricula and research programs, universities contribute to a widening gap between the skills demanded by technology-driven industries and the education they provide. As AI reshapes fields like healthcare, finance, logistics, and education, graduates from institutions that lack robust AI training find themselves at a disadvantage in the job market.

This divide disproportionately affects students from less privileged backgrounds who rely on public universities for career preparation. Without access to comprehensive AI education, these students are unable to compete with peers from private institutions or specialized programs that prioritize AI training. This technological inequity reinforces socioeconomic divides, limiting opportunities for upward mobility and exacerbating systemic disparities.

For example:

- ✓ Graduates entering the healthcare sector may lack the proficiency to work with AI-powered diagnostic tools.
- ✓ Students pursuing careers in logistics may be unprepared to manage AI-driven supply chain systems.

By failing to equip students with critical AI skills, universities leave them unprepared for the demands of the modern workforce, deepening inequality across generations.

Global Competitiveness: Falling Behind

Universities resistant to AI risk losing their relevance and influence on the global stage. Institutions that embrace AI are driving groundbreaking research, attracting top talent, and forging partnerships with industry leaders. These universities become innovation hubs, shaping policy, advancing economies, and contributing to societal progress.

In contrast, institutions clinging to outdated paradigms fall behind. They struggle to:

✓ Attract funding for AI-related research.
✓ Enroll international students drawn to cutting-edge programs.
✓ Build their global reputation as centers of innovation.

This decline not only weakens the universities but also impacts national progress. Countries with universities that fail to integrate AI into their education systems risk falling behind in global competitiveness, stifling economic growth and innovation on a broader scale.

Access and Inclusion: Marginalizing Underrepresented Communities

Resistance to AI adoption exacerbates disparities in education and access to opportunity, particularly for marginalized and underrepresented communities. Public universities often serve as a critical bridge for students from low-income or rural areas, offering a pathway to upward mobility through education. When these institutions fail to integrate AI into their curricula, they deny these students access to the tools and knowledge necessary to succeed in an AI-driven world.

This lack of inclusion not only limits individual opportunities but also perpetuates systemic inequities:

✓ **Underrepresentation in AI Development**: Without diverse voices in AI education, marginalized groups remain excluded from shaping AI technologies and policies.

✓ **Missed Opportunities for Social Mobility**: Students from disadvantaged backgrounds are left without the skills required to access high-demand AI-related jobs.

By resisting modernization, universities risk reinforcing societal barriers and failing to democratize the transformative potential of AI.

Universities as Barriers to Progress

Hesitation to adapt to the AI era positions universities as barriers to progress rather than catalysts for change. Their resistance perpetuates disparities in:

✓ **Education**: Failing to prepare students for AI-driven industries.

✓ **Employment**: Contributing to workforce imbalances and job market inequities.

✓ **Global Influence**: Undermining national and institutional competitiveness.

The ripple effects extend beyond campuses, affecting industries, governments, and societies that depend on universities for innovation and thought leadership.

The Need for a Paradigm Shift

Addressing these divides requires a fundamental shift in how universities approach AI. They must embrace their role as agents of transformation, fostering inclusion, equity, and innovation. This paradigm shift involves:

✓ **Modernizing Curricula**: Incorporating AI education, interdisciplinary learning, and ethical considerations to prepare graduates for a rapidly evolving workforce.

✓ **Industry Partnerships**: Collaborating with technology leaders to provide students with practical experience and access to cutting-edge tools.

✓ **Democratizing AI Education**: Ensuring that all students, regardless of background, have access to AI learning opportunities.

✓ **Fostering Diversity in AI Development**: Actively including marginalized voices in AI research and policymaking to create more equitable and representative technologies.

A Vision for Inclusive Progress

Universities have the potential to reclaim their leadership in shaping the future of AI by prioritizing innovation and equity. Resistance to AI adoption only deepens societal divides and limits the transformative potential of education. By modernizing their policies and practices, universities can ensure their graduates are not just participants but leaders in an AI-driven world.

This vision requires bold action and a commitment to bridging the divide between tradition and transformation. In doing so, universities can foster a future where AI serves as a tool for progress, inclusion, and opportunity, benefiting all of humanity rather than a privileged few.

8.3 Suggestions for Proactive Adaptation

To effectively navigate the rapidly evolving landscape of artificial intelligence (AI) and bridge the gap between tradition and modernity, universities must implement strategic, forward-thinking changes. These initiatives should equip students with essential skills for an AI-driven world while maintaining ethical integrity and fostering inclusivity. Below are actionable suggestions to transform universities into hubs of innovation, inclusivity, and societal progress.

1. Revise Curricula

The foundation of preparing students for an AI-driven future lies in modernizing university curricula. Traditional syllabi that emphasize theoretical concepts without practical application fail to meet the demands of today's workforce.

✓ **Practical Training**: Incorporate hands-on learning with AI tools such as TensorFlow, PyTorch, or data visualization platforms, enabling students to develop and deploy machine learning models.

✓ **Interdisciplinary Perspectives**: Introduce modules on ethics, societal impacts, and policy considerations to provide students with a comprehensive understanding of AI's broader implications.

✓ **Project-Based Learning**: Encourage students to tackle real-world problems using AI, fostering critical thinking and practical skills. For example, projects could focus on AI solutions for environmental sustainability or public health challenges.

By aligning academic programs with industry needs, universities can ensure that graduates are not just knowledgeable but also job-ready.

2. Foster Industry Collaboration

Collaborating with technology companies, startups, and research organizations bridges the gap between academic theory and practical application. Such partnerships benefit both students and faculty by providing access to cutting-edge tools, real-world challenges, and industry insights.

✓ **Internship Programs**: Partner with industry leaders to offer hands-on experience in fields like healthcare, sustainability, or finance.

✓ **Joint Research Initiatives**: Engage in collaborative projects that address pressing global challenges, such as developing ethical AI frameworks or AI-driven disaster response systems.

✓ **Trend Alignment**: Regular collaboration ensures that universities remain informed about emerging industry trends, enabling them to update curricula and research priorities accordingly.

This symbiotic relationship strengthens both academia and industry, fostering innovation and producing graduates equipped for the modern workforce.

3. Create AI Research Hubs

Establishing dedicated AI research centers positions universities as leaders in innovation, driving advancements in both theoretical and applied AI.

✓ **Interdisciplinary Collaboration**: Bring together experts from diverse fields such as computer science, ethics, sociology, and environmental science to explore AI's multifaceted impact.

✓ **Focus Areas**: Research hubs can address global challenges, such as improving urban infrastructure with AI or developing predictive models for climate resilience.

✓ **Funding and Partnerships**: Research centers can attract grants, talent, and industry collaborations, further solidifying the university's reputation as a pioneer in the AI era.

By fostering inquiry and innovation, these hubs serve as incubators for transformative solutions and thought leadership.

4. Implement Adaptive Policies

Rigid policies can hinder innovation, deterring researchers from pursuing ambitious projects. Universities must adopt adaptive frameworks that balance ethical safeguards with the need for experimentation.

✓ **Data Privacy**: Use advanced techniques like anonymization and secure sharing protocols to protect sensitive data while enabling meaningful research.

✓ **Regulatory Sandboxes**: Create controlled environments for testing AI applications, allowing researchers to explore risks and benefits responsibly.

✓ **Dynamic Oversight**: Continuously update policies to reflect advancements in AI, ensuring they remain relevant and supportive of innovation.

These measures foster a culture of responsible experimentation, empowering researchers to push boundaries while addressing ethical concerns.

5. Invest in Faculty Development

Faculty play a pivotal role in delivering quality AI education and driving research. Universities must ensure that educators remain at the forefront of AI advancements.

✓ **Professional Development**: Offer training programs, workshops, and access to the latest AI tools and methodologies.

✓ **Industry Engagement**: Encourage faculty to participate in collaborative projects or attend international conferences to enhance their expertise.

✓ **Incentivize Innovation**: Recognize and reward faculty contributions to AI research and curriculum development.

By equipping educators with cutting-edge knowledge, universities ensure that students receive instruction that is both current and relevant.

6. Emphasize Inclusion and Accessibility

AI education must be accessible to all, including underrepresented groups and non-traditional learners. Universities can democratize AI knowledge through the following initiatives:

✓ **Scholarships**: Provide financial assistance to students from marginalized communities to ensure equitable access.

✓ **Online Learning Platforms**: Offer affordable, flexible, and high-quality AI courses to reach diverse learners globally.

✓ **Community Outreach**: Organize workshops, hackathons, and awareness programs in underserved areas to inspire and educate future AI leaders.

By fostering inclusivity, universities can enrich AI development with diverse perspectives, leading to more equitable and innovative solutions.

Charting a Course for the Future

Adopting these proactive strategies enables universities to transform into dynamic centers of AI education and research. This evolution goes beyond keeping pace with technological advancements—it positions universities as catalysts for change, shaping a future where innovation and ethics coexist.

By integrating practical training, fostering collaboration, and prioritizing inclusivity, universities can prepare students not only to succeed but also to lead in the AI-driven world. These measures ensure that institutions uphold their responsibilities to equity, progress, and societal betterment while unlocking the transformative potential of artificial intelligence.

Building a Future-Ready Academia

Universities stand at a crucial crossroads, balancing their deep-rooted traditions with the pressing need to adapt to rapid technological advancements. As artificial intelligence (AI) reshapes industries, economies, and societies, academia faces a dual challenge: embracing innovation while upholding its foundational values. By addressing policy gaps, fostering inclusivity, and prioritizing innovation, universities can position themselves as leaders in an AI-driven future, equipping students to thrive in a world defined by constant change.

Reassessing Policies for Progress

University AI policies often lean toward risk minimization, which, while necessary, can inadvertently inhibit innovation. To prepare for the future, institutions must strike a balance between caution and creativity.

✓ **Flexible Data Usage Policies**: Enabling responsible access to vital datasets can catalyze advancements in critical areas such as healthcare, climate science, and education. Techniques like anonymization and secure sharing can address privacy concerns while fostering research.

✓ **Integrating Ethics and Practicality**: Academic frameworks must go beyond theory, embedding ethical considerations into real-world AI applications. This dual focus ensures that AI development aligns with societal values while fostering creativity and innovation.

By revising outdated policies, universities can create an environment where innovation thrives alongside accountability and ethical integrity.

Embracing Innovation as a Core Value

Innovation must become a cornerstone of academia. Universities should not only study AI but also harness its capabilities to enhance education, research, and operations.

✓ **AI-Driven Learning**: Leveraging AI to analyze learning patterns can enable personalized education tailored to individual needs, maximizing student potential.

✓ **Accelerating Research**: AI-powered tools can fast-track discoveries in fields such as renewable energy, medical treatments, and urban planning, addressing pressing global challenges.

✓ **Streamlining Campus Operations**: AI can optimize resource allocation, improve administrative efficiency, and enhance decision-making, allowing institutions to focus more on their academic mission.

By integrating AI across all aspects of university life, institutions can become models of technological adoption and innovation.

Fostering Inclusivity for Equitable Impact

Inclusivity must be central to academia's transformation. AI education and opportunities should be accessible to all, particularly to underrepresented and marginalized groups.

✓ **Scholarships and Financial Aid**: Providing support for students from underserved communities ensures equitable access to AI education.

✓ **Online Learning**: High-quality, affordable online AI courses can democratize knowledge, reaching learners regardless of their location.

✓ **Community Engagement**: Partnering with local organizations to host workshops and outreach programs can inspire and educate future AI leaders from diverse backgrounds.

Diversity within AI research teams enhances creativity, bringing varied perspectives that lead to more equitable and impactful solutions.

Leadership and Collaboration for Systemic Change

Preparing for an AI-driven future requires visionary leadership and strategic partnerships. Universities must champion bold initiatives that align traditional values with contemporary needs.

✓ **Revamping Curricula**: Academic leaders should prioritize updating outdated syllabi, incorporating interdisciplinary approaches and practical training with AI tools.

✓ **Establishing Research Hubs**: Interdisciplinary centers focused on AI innovation can drive breakthroughs while addressing ethical and societal concerns.

✓ **Fostering Public-Private Partnerships**: Collaborations with technology firms provide students with real-world experience, access to cutting-edge tools, and insights into industry trends.

These collaborative efforts ensure that academia remains relevant and continues to play a pivotal role in shaping technological progress.

Shaping Ethical and Practical AI Integration

Universities bear the dual responsibility of preparing students for the AI era and influencing the technology's ethical evolution.

- ✓ **Facilitating Dialogue**: Engaging diverse stakeholders—policymakers, technologists, and community representatives—ensures AI policies reflect societal values.
- ✓ **Investing in Ethical Research**: Universities should lead efforts to explore the ethical implications of AI, fostering transparency and accountability in its development.
- ✓ **Promoting Responsible Use**: By demonstrating how AI can address societal challenges while upholding integrity, universities can inspire confidence in the technology's potential.

This balanced approach enables academia to serve as both an innovator and a guardian of ethical standards in AI integration.

Charting the Path Forward

The intersection of tradition and transformation presents universities with a unique opportunity to redefine their role in society. By embracing innovation, addressing systemic shortcomings, and fostering inclusivity, academia can reestablish itself as a driving force of progress.

This evolution requires courage, adaptability, and a forward-thinking mindset—but the potential rewards are immense. Future-ready universities can prepare students to navigate a rapidly changing world, contribute to ethical AI development, and ensure that the benefits of technology are shared equitably across society.

By leading with vision and purpose, academia can shape a future where technology serves humanity responsibly and inclusively, ensuring progress and tradition coexist harmoniously.

Summary

This chapter critically evaluates how universities, as traditional bastions of knowledge and innovation, struggle to adapt to the transformative potential of AI and emerging technologies. It explores the disconnect between academic institutions' traditional frameworks and the rapidly evolving demands of the modern world, highlighting areas for improvement to ensure that universities remain relevant and impactful.

Key Themes:

Traditionalism vs. Modernization: Universities often operate within rigid, outdated frameworks that prioritize established disciplines over interdisciplinary and technology-driven approaches.

This resistance to change is partly due to institutional inertia and a reluctance to deviate from conventional pedagogical and research practices.

Gaps in Curriculum and Skills Development: Many universities fail to align their curricula with the skills needed in the AI-driven job market.

Students graduate with theoretical knowledge but lack practical expertise in applying AI and other technologies to real-world challenges.

Limited Adoption of AI in Academia: The use of AI tools for teaching, research, and administration is often restricted due to lack of awareness, skepticism, or resource constraints.

Institutions are missing opportunities to leverage AI for personalized learning, enhanced research capabilities, and operational efficiency.

Challenges of Ethical AI Integration: Universities struggle to incorporate ethical AI considerations into their programs, which are crucial for equipping students to navigate the moral complexities of technology.

There is insufficient focus on fostering a balanced understanding of AI's potential benefits and risks.

Siloed Approach to Knowledge: Academic departments often work in isolation, stifling the cross-disciplinary collaboration that is essential for AI innovation.

The lack of communication between fields like computer science, humanities, and social sciences limits the scope of creative problem-solving.

Need for Faculty Development: Many educators lack exposure to modern technological advancements, hindering their ability to prepare students for the machine age.

Professional development and AI literacy for faculty are critical to bridging this gap and fostering a culture of adaptability.

Proposed Pathways to Transformation: The chapter suggests actionable reforms, such as:

✓ Updating curricula to reflect industry trends and AI applications.

✓ Encouraging interdisciplinary research and learning.

✓ Establishing partnerships with industries to offer hands-on experiences for students.

✓ Leveraging AI to enhance teaching methodologies, improve administrative functions, and democratize access to knowledge.

Universities as Catalysts for Change: The chapter emphasizes the unique role universities can play as hubs of ethical innovation and societal progress.

By shedding "old lenses" and adopting forward-thinking approaches, institutions can lead the way in shaping the next generation of thinkers, creators, and leaders.

This chapter underscores the urgency for universities to rethink their roles and methodologies in light of technological advancements. It calls for a shift from traditionalism to a more dynamic, inclusive, and tech-savvy academic ecosystem that prepares students to thrive in a rapidly changing world.

Part IV: Redefining Creativity for the Future

"The future of creativity lies not in resisting change, but in embracing it with purpose—blending tradition with innovation to shape a world where imagination knows no limits."

— Sandeep Chavan

9: The Essence of Creativity

Creativity, at its core, transcends the instruments we use to express it. Whether shaped by paintbrushes or algorithms, it remains an innate human quality—a testament to our imagination, emotion, and intention. In the AI era, where tools can generate art, music, and writing with remarkable efficiency, the essence of creativity faces renewed scrutiny. What defines originality when machines contribute to the process? Is creativity rooted in the act of creation, the final product, or the purpose behind it? This chapter delves into these profound questions, exploring the timeless nature of creativity and its evolving relationship with the tools that amplify, but never replace, the human spirit.

9.1 Redefining Originality in the Age of AI

The advent of artificial intelligence (AI) in creative fields has sparked a fascinating debate about the nature of originality. As machines generate art, compose music, and craft stories, questions arise about whether these creations can be considered truly original. The concept of originality, once firmly rooted in human imagination, is being reexamined in an era where humans and machines collaborate to produce works that challenge traditional definitions of creativity.

Can Machines Be Truly Original?

At first glance, artificial intelligence (AI) systems like GPT and DALL-E seem to exhibit the ability to create original works. They generate intricate artworks, innovative designs, and compelling prose that often feel fresh and surprising, prompting the question of whether machines can truly innovate. However, the nature of AI's creative process reveals a more nuanced reality.

AI operates as a synthesis machine. It processes vast datasets, identifying patterns, combining elements, and producing outputs that mimic creativity.

Unlike humans, AI does not imagine, feel, or create from a blank slate. Its outputs are deeply rooted in the data it has been trained on, which is inherently human-made. For example, an AI trained on classical art might produce a painting that blends the styles of Van Gogh and Monet. While visually unique, the result is derivative, shaped entirely by pre-existing knowledge.

What AI lacks is the lived experiences, emotions, and subjective interpretation that define human creativity. Human originality is not just about producing something novel; it's about infusing creations with personal meaning, cultural context, and emotional resonance. A human artist painting about grief or hope channels their lived experiences, creating works that connect with audiences on a deeply personal level. In contrast, AI outputs, while aesthetically impressive, lack this emotional core and intentionality.

AI can emulate style and form but cannot grasp the broader significance of its creations. Its inability to reflect on its choices or imbue them with purpose further distinguishes it from human creativity. While AI can serve as a valuable tool for inspiration and exploration, the essence of originality—defined by depth, intentionality, and meaning—remains a distinctly human trait.

The Human Element: Emotion and Perspective

What truly sets human creativity apart from AI is the emotional depth and perspective embedded in every creation. Humans create not merely to solve problems but to express their inner worlds and grapple with complex emotions. Whether through art, music, literature, or design, these works carry personal stories and reflections on the human experience.

A poet writing about love or a painter capturing the fleeting beauty of a sunset infuses their work with emotions drawn from their lives. These creations are windows into the creator's soul, offering authenticity that resonates universally. The shared emotional journey between creator and audience becomes a defining feature of human artistry, fostering connection and meaning.

In contrast, AI operates without this emotional foundation. It can analyze patterns and replicate stylistic elements, but it cannot feel, reflect, or attach significance to its outputs. For example, an AI-generated painting may be visually stunning, but it lacks the vulnerability, intention, or story that a human artist brings to their work.

This difference is especially apparent in storytelling. Human storytellers draw from cultural, emotional, and personal contexts to weave narratives that inspire, challenge, and heal. AI-generated stories, though they may follow narrative structures and mimic human language, lack insight and genuine reflection.

The emotional and experiential core of human creativity underscores why originality remains uniquely human. AI serves as a tool to enhance and inspire, but it cannot replicate the depth that arises from human introspection and emotion.

Collaboration or Competition: AI as a Partner in Creation

The relationship between humans and AI in creative fields is best viewed as a collaboration rather than a competition. AI offers tools to enhance human creativity, pushing boundaries and opening new possibilities. This partnership allows creators to achieve results that might have been impossible otherwise.

AI excels at analyzing vast amounts of data, identifying patterns, and generating suggestions. These strengths complement human creativity, providing a foundation upon which artists, designers, and innovators can build. For instance:

✓ **In Architecture**: AI can generate multiple design iterations, optimizing for environmental factors and aesthetic preferences. Architects can refine these outputs, blending precision with artistic vision.

✓ **In Music**: AI-generated melodies inspire musicians to explore new styles, but it is the musician who imbues the composition with emotional depth and cultural context.

✓ **In Writing**: Writers use AI tools to brainstorm ideas or refine language, maintaining control over tone, intent, and narrative direction.

This dynamic interplay creates a new form of originality, emerging not from the machine or human alone but from their interaction. While the human creator defines the vision and interprets results, AI provides computational power, efficiency, and fresh perspectives.

However, this collaboration raises questions about authorship. In works created through human-AI partnerships, who deserves credit? The human creator drives intent and meaning, yet the AI's contributions cannot be ignored. Rethinking authorship acknowledges the collaborative nature of modern creativity.

By viewing AI as a partner, creators can expand their horizons, redefining the essence of artistic and intellectual exploration. This collaborative approach amplifies human ingenuity, transforming how we create and innovate.

Redefining Originality in an AI Era

As AI continues to evolve, it challenges traditional notions of originality. Machines assist, inspire, and amplify human creativity, but the essence of originality—emotional depth, perspective, and purpose—remains deeply human.

In the AI era, originality is best understood as a spectrum of contributions. Machines provide tools to explore new possibilities, while humans infuse creations with meaning and resonance. Together, they push the boundaries of what is possible, blending technological innovation with the enduring beauty of human expression.

Ultimately, the future of creativity lies not in machines replacing humans but in their collaboration, ensuring that the stories, art, and ideas we create continue to reflect the diversity, complexity, and depth of the human spirit.

9.2 The Multifaceted Nature of Creativity

Creativity is a concept with many layers, each offering a unique lens through which to understand the act of creation. It is not confined to a single definition or dimension; instead, it encompasses a spectrum of experiences, outcomes, and purposes. By examining creativity as a process, an outcome, and an intention, we can better appreciate its complexity and significance in human expression.

Creativity as Process: The Journey of Expression

At its core, creativity is a journey—a dynamic and evolving process of exploration, experimentation, and discovery. It is less about reaching a predefined goal and more about immersing oneself in the act of creating. This perspective shifts focus from the outcome to the unfolding of ideas, the iterative refinement of techniques, and the unpredictable transformation of thoughts into reality. The process is deeply personal, allowing creators to connect with their inner selves and translate abstract emotions into tangible forms.

For many, the magic of creativity lies in the process itself. A painter revels in blending colors and textures, witnessing a blank canvas transform with each stroke. A musician discovers uncharted emotional landscapes as they experiment with melodies and rhythms. A writer, navigating the twists of a developing story, uncovers truths about their characters—and themselves—that they hadn't anticipated. These moments of engagement bring profound satisfaction, often eclipsing the gratification of completing the final product.

Focusing on the process cultivates curiosity and openness, encouraging creators to embrace uncertainty and take risks. Mistakes and unexpected outcomes are reframed not as failures but as integral steps in the journey. A sculptor might accidentally chip away too much material, only to realize the resulting form better captures their vision. A researcher, surprised by unforeseen results, may uncover entirely new avenues of inquiry. These serendipitous moments highlight creativity's iterative nature, where every action—whether deliberate or accidental—adds depth and dimension to the work.

Engaging in the creative process also fosters resilience. Facing challenges, doubts, and setbacks becomes an opportunity for growth, honing problem-solving skills, adaptability, and confidence. This resilience transcends the creative act, equipping individuals with a mindset that can be applied to life's broader challenges.

For many, creativity is also therapeutic—a means of self-expression and emotional release. A dancer channels grief or joy into movement, while a photographer captures fleeting moments that evoke introspection and connection. These acts of creation are as much about the creator's internal journey as they are about the external product.

Ultimately, embracing creativity as a process reframes creation as a source of joy, discovery, and transformation. It reminds us that success lies not solely in the destination—the finished work—but in the experience itself. The challenges faced, insights gained, and growth achieved along the way are the true rewards.

Creativity as Outcome: Valuing the Final Product

While the creative journey is deeply fulfilling, the outcome—the tangible product of that journey—holds its own unique significance. The final work embodies the creator's vision, skill, and perseverance, serving as a bridge between creator and audience. It transforms abstract ideas into forms that can be shared, experienced, and appreciated.

The outcome often defines the broader impact of creativity. A scientific breakthrough might revolutionize an industry, a novel might ignite introspection, or an architectural masterpiece might inspire generations. These creations leave a lasting mark on society, advancing knowledge, shaping culture, and altering perceptions. The final product is celebrated as a testament to human ingenuity, offering both practical solutions and aesthetic enrichment.

Beyond its immediate purpose, the outcome becomes a legacy. A symphony may encapsulate the emotions of its era, resonating with audiences long after its composer has passed. An innovative technology symbolizes progress, showcasing how imagination and determination drive humanity forward.

Valuing the outcome complements the importance of the process, highlighting their interconnectedness. The final product is the culmination of countless iterations, choices, and inspirations that occurred during the creative journey. It is a visible representation of the creator's ability to overcome challenges, refine ideas, and bring their vision to life.

The outcome also serves as a medium of connection and dialogue. A powerful film might spark conversations about societal issues, while a beautifully designed space fosters a sense of comfort and belonging. The audience's interaction with the final work breathes new life into it, adding layers of interpretation and meaning that transcend the creator's original intent.

Moreover, the final product inspires future creativity. An invention or artistic work can push others to explore new possibilities, fueling the evolution of ideas. Creativity as outcome reminds us that the act of creation is not complete until the work is shared and experienced. Whether it moves, challenges, or uplifts, the final product celebrates the transformative power of human imagination.

Creativity as Intention: The Purpose Behind Creation

Beyond the act of creating and its outcome lies a powerful undercurrent: intention. Intention is the "why" behind the work, infusing it with meaning and purpose. It reflects the creator's values, aspirations, and worldview, elevating creativity from a mechanical task to a deliberate and meaningful act.

For some, creativity is a form of self-expression—a way to communicate emotions, share personal experiences, or tell resonant stories. A painter may channel grief into a poignant portrait, while a writer crafts narratives to explore universal truths about love or resilience. These works resonate because they stem from authentic human experiences.

For others, creativity is a means to solve problems or spark change. A scientist might invent a device to address pressing challenges, or an activist might use art to amplify marginalized voices. Here, intention is outward-facing, aiming to create tangible societal impact. The purpose drives the creator to innovate, contribute, and transform.

Intention also shapes how audiences interact with creative works. Understanding the purpose behind a piece provides context, enriching its interpretation. A painting created in response to personal loss gains emotional weight, while a technological breakthrough becomes more profound when linked to its potential to solve global issues. Intention bridges the gap between creator and audience, fostering empathy and connection.

Intention influences the creative process as well. A creator motivated by curiosity might approach their work playfully, while one driven by social justice might focus on urgency and impact. Intention ensures that every element—materials, methods, or themes—aligns with the creator's purpose, imbuing the work with coherence and meaning.

Intention also raises questions of responsibility. With creativity comes the power to influence thoughts, emotions, and actions. Creators must consider the ethical implications of their work, ensuring it contributes positively to society. This reflection transforms creativity into an intentional act of shaping not just art or ideas but the broader human experience.

Ultimately, intention distinguishes human creativity from machine-generated outputs. While AI can replicate processes and produce outcomes, it cannot embed them with the personal meaning, cultural relevance, or ethical considerations that define human originality.

The Synergy of Process, Outcome, and Intention

Creativity is not confined to one dimension—it is a symbiotic blend of process, outcome, and intention. The process brings joy and discovery, the outcome creates impact and legacy, and intention imbues both with depth and purpose. Together, these elements make creativity a transformative force in personal and collective experiences.

By recognizing and embracing this synergy, creators can find fulfillment in every stage of their journey, ensuring that their work resonates, inspires, and endures. In a world where technology and human ingenuity increasingly intersect, the essence of creativity—rooted in humanity's unique ability to imbue work with meaning—remains a timeless testament to our shared spirit and vision.

9.3 Originality vs. Influence: Where the Line Blurs

The concept of originality has long been a cornerstone of creativity, celebrated as the ability to produce something entirely new. However, creativity rarely emerges in a vacuum; it is often shaped by external influences, past experiences, and cultural narratives. This interplay between originality and influence becomes even more complex in the age of AI, where algorithms trained on vast datasets blur the line between inspiration and replication. Understanding this dynamic requires exploring the role of inspiration in original work, the challenges posed by AI, and strategies for maintaining authenticity in an increasingly digital world.

The Role of Inspiration in Original Work

Inspiration is the cornerstone of creativity, the spark that ignites the creative process. It provides the raw materials from which original ideas are sculpted, offering a vital connection between the creator and the world around them. Far from existing in isolation, inspiration arises from interactions with art, nature, culture, personal experiences, or the works of others. These influences are not merely incidental but integral, shaping the context and depth that allow new ideas to flourish.

A musician might draw from classical symphonies, infusing their complexity and emotion into a contemporary composition that resonates with modern audiences. A novelist could weave historical events into fictional narratives that reveal universal truths about human nature. These influences enrich the creative process, layering meaning into the final work and creating connections between creator and audience. Rather than diminishing originality, they act as catalysts, enabling creators to reinterpret and reimagine existing ideas in transformative ways.

Originality, therefore, does not spring from isolation or novelty for its own sake. It emerges from the transformation of existing elements into something uniquely personal. A painter inspired by a sunset might translate its hues and atmosphere into a visual language reflecting their emotions. A filmmaker influenced by a specific genre might subvert its conventions to tell a story entirely their own. This process of reinterpretation and personalization distinguishes genuine originality from mere replication.

However, the boundary between inspiration and imitation can be delicate. Creators must navigate this line carefully, ensuring their work retains a distinct voice and vision. While it is natural and even necessary to draw from external sources, true originality lies in internalizing those influences and expressing them through one's unique perspective. This approach transforms borrowed elements into authentic, resonant creations.

Inspiration also highlights the interconnectedness of creativity. Each idea builds upon what came before, forming an ever-evolving tapestry of human expression. This interplay between influence and originality keeps creativity dynamic and expansive, offering infinite possibilities for reinterpretation and growth.

Ultimately, inspiration is not a limitation but a source of empowerment. By engaging with the world and its myriad influences, creators expand their horizons, explore new ideas, and craft works that are deeply personal yet universally resonant. The act of transforming inspiration into originality defines the creative spirit, allowing it to thrive in a constantly evolving landscape.

How AI Challenges Traditional Notions of Influence

The rise of artificial intelligence (AI) has revolutionized how we understand influence and originality in creativity. Trained on vast datasets of human-generated content, AI systems like GPT and DALL-E generate outputs that often appear innovative and original. These models produce poetry, art, music, and other creative works that mimic originality, raising questions about what it means to create something new.

Unlike traditional tools that execute tasks based on explicit instructions, AI systems analyze, learn, and synthesize at an unprecedented scale and speed. They blend artistic styles, reinterpret genres, and generate novel concepts by leveraging patterns in their training data. This creates a paradox: AI outputs can seem fresh and surprising, yet they are inherently derivative, deeply rooted in pre-existing material.

This dynamic complicates the distinction between inspiration and replication. For instance, an AI-generated painting might combine the techniques of Van Gogh and Monet, producing a visually striking and seemingly original piece. However, the originality is limited by the fact that it relies on these influences, raising ethical questions about authorship and intellectual property.

AI also democratizes creativity, enabling individuals from diverse backgrounds to experiment with styles and techniques that might otherwise be inaccessible. This democratization broadens creative horizons but introduces the risk of homogenization. If creators rely heavily on similar AI tools, the diversity of creative outputs may diminish, leading to a convergence of styles.

Despite these challenges, AI offers significant opportunities for innovation. By presenting unexpected combinations and interpretations, it can inspire human creators to push boundaries and explore uncharted territories. An architect might use AI to visualize unconventional designs, or a musician might draw inspiration from AI-generated melodies that spark fresh ideas. In this collaborative context, AI becomes a partner in the creative process, enhancing human ingenuity rather than replacing it.

However, the integration of AI into creativity necessitates a rethinking of authorship and originality. Creators must ensure that their unique perspective and intent remain central, using AI as a tool to amplify their vision rather than overshadow it. This evolving relationship challenges traditional notions of influence while opening new pathways for collaboration and exploration.

Maintaining Authenticity in a Digital Landscape

In today's interconnected and technology-driven world, maintaining authenticity is more critical than ever. Authenticity separates original work from imitation, anchoring it in the creator's unique vision, voice, and values. It is this authenticity that allows creative works to resonate deeply, forging connections that transcend surface-level appeal and speak to shared human experiences.

The rise of AI-powered tools has transformed the creative process, offering unprecedented possibilities for innovation. However, these advancements blur the line between authentic creation and algorithmic synthesis. To preserve authenticity, creators must take an intentional approach to their work, ensuring that their personal voice remains distinct and central.

Creators can uphold authenticity by engaging thoughtfully with digital tools. Rather than relying on AI to replace creativity, they can use it as a collaborator. For example, a writer might use AI to generate ideas or refine prose but ensure that the final narrative reflects their unique insights and emotions. Similarly, a visual artist might incorporate AI-generated designs into their work but infuse the finished piece with their personal style and meaning.

Transparency is another critical aspect of authenticity. Acknowledging the role of tools and influences—whether human or algorithmic—not only builds trust but also deepens the creative process. When creators openly share how they use AI or draw inspiration from other sources, they foster a greater appreciation for the effort and thought behind their work.

Authenticity also depends on the creator's ability to connect with their audience. Original work reflects individuality while resonating with universal emotions and experiences. This balance between personal expression and broader relevance ensures that creativity remains impactful and meaningful.

As digital tools continue to evolve, maintaining authenticity requires creators to go beyond technical skill and focus on emotional and intellectual depth. AI should enhance, not overshadow, the human element, preserving the creator's intent and perspective as the heart of the work. By doing so, creators safeguard their individuality and contribute to a culture of meaningful, ethical creativity.

The Balance Between Originality and Influence

The relationship between originality and influence is symbiotic, not oppositional. All creativity is shaped by what has come before, drawing from cultural, historical, and personal contexts. Originality lies in how these influences are transformed and reimagined, creating works that reflect the creator's unique perspective.

In the digital age, this balance has become more nuanced. AI systems add new dimensions to influence and originality, offering tools that process immense datasets and generate ideas at scale. While these tools provide unprecedented opportunities, they also challenge creators to maintain their distinct voice and vision.

Originality does not require rejecting influence but embracing it as a foundation for innovation. A painter inspired by Impressionism might reinterpret its principles to explore modern themes, or a composer influenced by folk traditions might integrate them into contemporary symphonies. What distinguishes original work is the creator's ability to internalize and reshape these influences, infusing them with their unique perspective.

The integration of AI into creativity raises new questions about authorship and authenticity. AI-generated outputs blur the line between inspiration and replication, making it essential for creators to use these tools thoughtfully. By maintaining their personal imprint and ensuring their intent remains central, creators can navigate this evolving landscape while preserving the authenticity that defines originality.

The balance between originality and influence underscores the dynamic nature of creativity. By honoring the past while pushing boundaries, creators craft works that are both grounded and forward-looking. This interplay ensures that creativity remains vibrant, relevant, and deeply human, even in an era of rapid technological advancement.

9.4 The Human Spirit: The Heartbeat of True Creativity

Creativity is more than a skill or a process; it is a profound reflection of the human spirit. At its core lies an innate desire to express, connect, and innovate. This essence of creativity transcends tools, techniques, or technologies, grounding itself in the unique qualities of human emotion, imagination, and values. The human spirit is the heartbeat of true creativity, the element that ensures every creation carries meaning and depth.

Why Emotion and Context Matter

Emotion and context are the lifeblood of human creativity, elevating it beyond technical skill to a profound form of expression that resonates deeply with others. These elements transform creative works into vessels of meaning, bridging personal experience and universal understanding. Together, emotion and context ensure that creativity not only captures the essence of its time but also speaks to timeless human experiences.

The Role of Emotion in Creativity

Emotion is the driving force that gives creative works their soul, allowing them to evoke responses that transcend logic. A song that captures the vulnerability of heartbreak or the joy of love speaks directly to the human condition, forging connections between the artist and their audience. These works serve as mirrors, reflecting shared emotions and creating a dialogue between creator and viewer.

When creators channel their emotions into their work, they imbue it with authenticity and depth. This emotional resonance allows individuals from diverse backgrounds to connect with the work on a personal level. A poem about resilience, for instance, can inspire and comfort someone facing struggles, regardless of cultural or linguistic barriers. Emotions make art universal, fostering empathy and shared understanding.

The Influence of Context on Creative Impact

While emotion provides depth, context gives creativity its relevance. Creative works are shaped by the time, place, and circumstances of their creation, anchoring them in the realities of their era. A novel written during a war may explore themes of loss and survival, while a painting born out of a cultural movement may reflect a collective yearning for change. Context lends significance to creative works, grounding them in the lived experiences of their time.

Moreover, context shapes how audiences interpret creative works. Picasso's *Guernica*, for example, is not just a masterful piece of art—it is a searing commentary on the horrors of war, made powerful by its connection to the Spanish Civil War. Without understanding the historical backdrop, much of the work's impact would be diminished. Context transforms art into a bridge between the creator's world and the audience's understanding, enriching its meaning and impact.

The Synergy of Emotion and Context

When emotion and context intersect, creativity reaches its full potential. Consider a filmmaker addressing contemporary social issues through an emotionally charged narrative. The emotional weight of the story draws viewers in, while the contextual relevance sparks conversations about the world they inhabit. This synergy amplifies the work's resonance, making it both deeply personal and widely impactful.

For creators, this interplay enriches the creative process. Emotion drives them to articulate their inner truths, while context challenges them to engage with the external world. Together, these forces shape works that are not only authentic to the creator's perspective but also relevant to their time and meaningful across generations.

Emotion and context give creativity its enduring power. They ensure that works are not just seen or heard but deeply felt and understood. By grounding their creations in these elements, creators forge connections that transcend cultural and temporal boundaries, leaving legacies that continue to inspire.

Imagination as the Gateway to Originality

Imagination is the wellspring of creativity, the faculty that enables us to envision what does not yet exist. It transforms what *is* into what *could be*, pushing the boundaries of possibility and driving originality. More than a tool for innovation, imagination is the essence of human creativity, allowing us to dream, explore, and create works that challenge and inspire.

The Role of Imagination in Creation

Imagination is the bridge between reality and possibility. A scientist might use it to hypothesize groundbreaking theories, while a poet crafts metaphors that capture ineffable truths. An inventor relies on imagination to design solutions that improve lives or disrupt industries. Across disciplines, imagination allows creators to challenge the status quo and envision new paradigms.

Unlike replication or adaptation, imagination reshapes the familiar into something entirely new. A novelist may take inspiration from historical events but weave them into a narrative that transcends time, offering fresh insights into human nature. A painter might reimagine the natural world through abstract forms, creating works that evoke emotions beyond what is visible. Imagination breathes originality into every creative act.

The Human Element: Imagination as a Unique Force

What sets human imagination apart is its unpredictability and depth. Unlike AI, which synthesizes data patterns, human imagination is unbounded, drawing from personal experiences, emotions, and subconscious insights. A musician might hear a melody in their mind, born of dreams or emotions, while an artist visualizes colors and forms inspired by fleeting moments of inspiration.

This spontaneity defines human originality. While AI can mimic styles and generate outputs that resemble creativity, it lacks the intuitive leaps and emotional depth that characterize human imagination. The meaning imbued in human creations often stems from connections that are deeply personal, irrational, and profoundly transformative—qualities machines cannot replicate.

Imagination as a Catalyst for Progress

Imagination is not only a source of artistic innovation but also a driver of progress. Visionaries like Nikola Tesla and Leonardo da Vinci used their imagination to bridge disciplines, creating technologies and ideas that revolutionized their fields. Imagination propels humanity forward, inspiring exploration, discovery, and growth.

In culture and art, imagination challenges norms and envisions new possibilities. A dystopian novel may provoke reflection on societal flaws, while a utopian painting sparks aspirations for a better future. Imagination fuels dialogue and transformation, shaping the way we understand ourselves and the world around us.

Imagination in a Digital Age

In an era dominated by AI, the significance of human imagination is magnified. While technology offers tools to assist and enhance creativity, it cannot replace the uniquely human ability to dream and reflect. A filmmaker might use AI for storyboarding but relies on their imagination to craft a narrative that resonates emotionally. Similarly, a designer might explore

AI-generated patterns but ultimately determines what aligns with their vision and identity.

Imagination ensures that creativity remains dynamic, transformative, and deeply personal, even in a landscape shaped by technology. It is the force that allows us to transcend the limitations of the present and explore the limitless possibilities of the future.

Creativity as a Reflection of Humanity's Values and Aspirations

Creativity is more than an act of self-expression—it is a profound reflection of humanity's collective soul. Through creative works, individuals and communities articulate their values, aspirations, and ideals, creating legacies that define their time and inspire future generations.

Creative Works as Cultural Mirrors

Creative works capture the spirit of their era, distilling its challenges, hopes, and dreams into tangible forms. A novel addressing inequality may give voice to marginalized perspectives, while a groundbreaking invention symbolizes human ingenuity and perseverance. These works not only respond to their time but also shape it, influencing thought, culture, and progress.

From Renaissance masterpieces to modern innovations, creative works resonate across time because they reflect universal truths. Beethoven's symphonies, for instance, echo the revolutionary fervor of his age, while Van Gogh's paintings convey the timeless beauty and anguish of the human condition. These creations endure because they speak to shared experiences and values.

Creativity as a Collective Story

Through creativity, humanity tells its collective story. Myths, legends, and cultural narratives preserve identities and convey moral lessons, serving as frameworks for understanding the world. Modern storytelling—through films, novels, or technology—continues this tradition, reflecting contemporary concerns while inspiring action and dialogue.

The Universality of Creativity

The most enduring works of creativity transcend their immediate context, resonating with people across cultures and generations. Whether it is a song that evokes shared emotions or a technological breakthrough that transforms lives, creativity bridges divides and highlights our common humanity.

Emotion, imagination, and shared values make creativity a powerful force for connection and progress. In an age where technology offers new tools, the essence of creativity remains firmly rooted in the human spirit. By embracing these elements, creators ensure their work continues to inspire, challenge, and unite us in a constantly changing world.

9.5 A New Paradigm: Blending Tools with Timeless Creativity

The intersection of advanced tools and human creativity marks the dawn of a new paradigm. While traditional creativity was often defined by manual effort and innate imagination, the integration of AI and other technologies has expanded the horizons of what creators can achieve. This chapter explores how tools can be harmoniously blended with timeless principles of creativity, ensuring that innovation enhances rather than diminishes the authenticity of the creative process.

Rethinking Tools as Extensions, Not Replacements

Tools have always been fundamental to human progress, amplifying our abilities rather than substituting our ingenuity. From the chisels and brushes of ancient artisans to the sophisticated algorithms of today, tools extend human creativity and open new dimensions of possibility. The advent of AI represents a transformative moment in this continuum, not as a threat to human creativity but as an opportunity to expand its horizons.

Tools as Amplifiers of Human Potential

At their best, tools reduce barriers, enabling creators to experiment and execute ideas with unprecedented ease and precision. AI exemplifies this by tackling tasks once considered too complex or time-consuming. A filmmaker can use AI to storyboard scenes with intricate detail, while a fashion designer might experiment with patterns that combine traditional aesthetics and modern trends.

Importantly, these tools do not replace the human touch. Instead, they empower creators to explore new dimensions of their craft. An AI-generated architectural concept, for example, may inspire innovative design, but it is the architect's personal insight and experience that guide the final realization. Similarly, musicians can use AI to compose harmonies, yet the emotional depth and cultural resonance of the piece depend on the human creator's intent and interpretation.

Creativity as Decision-Making

The essence of creativity lies in the decisions that shape the work, not merely in its execution. Tools like AI can generate options, streamline processes, and suggest variations, but they cannot replicate the nuanced decision-making that defines human creativity.

Consider a graphic designer who uses AI to explore color palettes. While the tool may offer endless combinations, the designer's choice of hues, guided by their understanding of mood, audience, and context, defines the outcome. By placing creative decision-making firmly in human hands, tools become collaborators rather than competitors, enhancing the creative process while preserving its authenticity.

Transforming Fear Into Opportunity

The fear that tools, particularly AI, might displace human creativity often stems from a misunderstanding of their role. Rather than replacing creators, tools expand the boundaries of what is possible, enabling new forms of expression and innovation.

A writer, for instance, might use AI to refine syntax or brainstorm plotlines, but the heart of the narrative—its themes, characters, and emotional arcs—remains firmly rooted in the writer's imagination and intent. This partnership between human creativity and machine capability not only enhances productivity but also pushes the creative process into uncharted territories.

A Collaborative Future

By viewing tools as partners, creators can harness their potential while maintaining control over the creative process. This collaborative approach leverages the strengths of both human and machine, achieving results neither could accomplish alone. Tools transition from being perceived as replacements to becoming enablers, celebrated for their ability to expand the scope of human ingenuity.

This paradigm shift transforms the narrative from fear to opportunity, positioning tools like AI as catalysts for innovation. When thoughtfully integrated, these tools amplify human creativity, fostering a symbiotic relationship that enhances both efficiency and artistry.

Preserving Intentionality in AI-Assisted Creation

As tools like AI become more integrated into creative workflows, preserving intentionality is paramount. Creativity is more than production; it is an expression of purpose. Without clear intent, even the most technically impressive works risk losing their authenticity and emotional resonance.

Purpose as the Core of Creativity

Intentionality transforms ideas into meaningful creations. When guided by purpose, every creative decision aligns with the creator's vision, ensuring authenticity. Tools like AI can support this process, but only when the creator remains in control. For example, an artist might use AI to generate visual concepts but must curate and refine the outputs, infusing them with personal perspective and emotional depth.

The Refinement Process

AI outputs often require human refinement to elevate them from functional to transformative. A writer using AI to brainstorm ideas, for instance, must adapt and shape those suggestions, aligning them with their narrative's tone and themes. This stage of refinement showcases the human element, where creators inject their experiences, values, and instincts into the work.

Transparency and Trust

Transparency about the role of tools in creation builds trust with audiences and enriches the creative process. Acknowledging the use of AI does not diminish the creator's contribution but instead positions the technology as an integral partner. This openness fosters a cultural shift, redefining AI as a complement to human ingenuity rather than a replacement.

Creativity in a Machine-Enhanced World

As technology continues to evolve, creativity is entering an era where the boundaries of possibility are being redefined. Advanced tools like AI, augmented reality, and design software enable creators to visualize, experiment, and execute ideas with unprecedented scope.

Expanding Creative Boundaries

These technologies empower creators to explore new mediums and tackle complex concepts. Architects can simulate entire environments in real time, while filmmakers craft immersive worlds using virtual reality. Yet the defining factor remains the creator's vision and intent, which give life and meaning to these possibilities.

Symbiosis: Human and Machine Collaboration

The future of creativity lies in symbiosis, where humans and machines collaborate seamlessly. Tools will handle repetitive tasks and offer innovative suggestions, freeing creators to focus on higher-level thinking and emotional resonance. Musicians might use AI to draft harmonies, but the emotional and cultural layers that define the music will always stem from human interpretation.

Democratizing Creativity

AI has the potential to democratize creativity, making tools accessible to individuals from diverse backgrounds. Aspiring filmmakers can edit videos with affordable software, and writers can refine their ideas using AI-powered platforms. However, democratization must be accompanied by ethical practices to ensure technology empowers rather than exploits.

Redefining Creativity for the Modern Age

The challenge is not to resist technology but to integrate it thoughtfully, blending timeless creative principles—imagination, authenticity, and intent—with cutting-edge innovation. By viewing tools as extensions of human capability, creators can preserve the integrity of the creative process while exploring the opportunities of a machine-enhanced world.

This approach ensures that creativity remains deeply rooted in human values and aspirations. It celebrates the fusion of human and machine, not as a threat to originality but as an elevation of what creativity can achieve.

Together, tools and creators can shape a future where innovation thrives without compromising the timeless essence of artistry and ingenuity.

Summary

This chapter delves into the fundamental nature of creativity, emphasizing its deep-rooted connection to human emotion, imagination, and cultural context. It argues that while tools, including AI, can enhance creative processes, the core of creativity remains a distinctly human endeavor. The chapter explores how creators can transcend reliance on tools to nurture originality and authenticity, ensuring that their work resonates on a deeper, more meaningful level.

Key Themes:

Creativity as an Innate Human Trait: Creativity is framed as an intrinsic human ability, shaped by personal experiences, emotions, and cultural narratives.

Tools, while valuable, are secondary to the creator's vision, serving as enablers rather than the source of creativity.

The Risk of Over-Reliance on Tools: Dependence on AI or other tools risks homogenizing creativity, potentially leading to works that lack individuality and emotional depth.

Tools can generate iterations, but they lack the intuition, empathy, and unpredictability that drive human originality.

Creativity Rooted in Emotion and Storytelling: True creativity often stems from the ability to convey complex emotions and stories that resonate universally.

For example, a painter's brush strokes or a writer's narrative voice carry a personal touch that cannot be replicated by machines.

Cultural Context and Personal Expression: Creative works are deeply influenced by cultural heritage and personal identity.

Creators who ground their work in their cultural or personal experiences bring authenticity and relevance that transcend trends and technological interventions.

Balancing Innovation with Timeless Principles: The chapter argues for a balanced approach: leveraging tools to enhance creativity without compromising the essence of one's craft.

For instance, while an AI model might help a musician create complex harmonies, the emotional connection of the piece lies in the human artist's intent and expression.

The Role of Reflection and Originality: Reflection on one's values, experiences, and inspirations fosters originality, enabling creators to explore new dimensions of their craft.

The chapter encourages creators to occasionally step away from tools to reconnect with their core creative impulses.

Reclaiming the Soul of Creativity: The chapter advocates for creators to view tools as collaborators rather than leaders in the creative process.

By focusing on imagination, innovation, and authenticity, creators can ensure that their work stands out in a machine-dominated world.

Creativity's Universal Essence: The universality of creativity lies in its ability to bridge gaps between individuals, cultures, and generations.

Works that are deeply human often achieve timeless relevance, inspiring audiences long after trends and technologies have faded.

This chapter champions the enduring power of human creativity, asserting that it is driven not by tools but by the creator's passion, imagination, and ability to connect with others. It encourages creators to embrace tools thoughtfully while keeping their focus on the essence of their craft—authentic expression and impactful storytelling.

10: Humans and Machines

The future of creativity and innovation lies in collaboration—not just between individuals, but between humans and the machines they build. As artificial intelligence and automation become increasingly integrated into our daily lives, the question is no longer whether humans and machines will coexist, but how effectively they can work together. This collaborative future holds immense promise, with AI amplifying human potential and humans providing the emotional depth, ethical oversight, and creativity that machines lack. By embracing this partnership, we can navigate the complexities of a rapidly evolving world, ensuring that technology enhances, rather than diminishes, the uniquely human qualities that drive progress.

10.1 The Power of Human Adaptability in a Tech-Driven World

Adaptability is the cornerstone of human progress, enabling us to thrive amidst waves of technological change and societal evolution. From mastering fire to building complex digital networks, humanity has consistently transformed challenges into opportunities. In today's tech-driven world, where AI and automation are reshaping industries and lives, this capacity for adaptation remains as critical as ever. By embracing change and leveraging our inherent flexibility, we not only navigate disruption but also uncover new possibilities for innovation and growth.

Embracing Change: Lessons from History

Human history is a testament to the transformative power of adaptation. From the discovery of fire to the rise of the digital age, humans have continually reshaped their environments and redefined what is possible. This adaptability has enabled societies to not only survive disruptions but to thrive in the face of them.

Consider the agricultural revolution, which shifted humans from nomadic lifestyles to settled farming communities. This change required a rethinking of humanity's relationship with nature, the development of tools for cultivation, and the organization of societies around food production. Similarly, the industrial revolution introduced mechanized processes that transformed industries, economies, and daily life. While these changes initially raised fears about job displacement and societal upheaval, they ultimately fueled unprecedented growth and innovation.

The invention of the printing press provides another compelling example. Initially resisted by traditional knowledge gatekeepers, it democratized access to information and catalyzed transformative movements like the Renaissance, the Enlightenment, and the development of modern education. These milestones illustrate a recurring theme: resistance to change often gives way to acceptance and progress, underscoring humanity's resilience and ability to harness new tools for advancement.

In today's era of AI and automation, similar fears and uncertainties have emerged. However, as history demonstrates, the human capacity for adaptation remains pivotal. By embracing change and learning from past experiences, societies can navigate the complexities of technological advancements while safeguarding cultural and ethical values.

Flexibility as a Hallmark of Human Evolution

Flexibility has been a defining characteristic of humanity, enabling adaptation to unpredictable circumstances and fostering creative problem-solving. Across millennia, humans have shown remarkable ability to adjust to new environments and challenges, from migrating across continents during the Ice Age to integrating digital tools into contemporary workplaces.

This adaptability is powered by uniquely human traits such as curiosity, emotional intelligence, and the ability to learn from diverse experiences. Unlike machines, which operate within predefined parameters, humans can pivot when faced with unforeseen challenges, using intuition and imagination to chart new paths. For example, the rapid shift to remote work technologies during the COVID-19 pandemic showcased this flexibility. Individuals and organizations quickly adopted digital platforms, redefining collaboration and productivity amid global disruption.

Flexibility also plays a critical role in driving innovation. Engineers, for instance, often encounter unexpected outcomes during development, leading to breakthroughs beyond their original objectives. Similarly, AI technologies initially designed for specific applications have been repurposed across medicine, education, and entertainment, thanks to human ingenuity.

In a technology-driven world, this blend of adaptability and creativity ensures that humans remain indispensable. While AI excels at data analysis and executing complex tasks, it lacks abstract thinking, empathy, and moral reasoning—qualities that drive innovation and anchor technology in ethical and meaningful applications.

The Driving Force of Adaptability

Adaptability is not merely a survival mechanism; it is the engine of progress. Throughout history, humans have consistently turned challenges into opportunities by embracing change and leveraging flexibility. As AI and automation reshape the world, this adaptability will be vital in ensuring that technology complements, rather than replaces, human potential.

The challenge is not to resist change but to approach it with curiosity, resilience, and a willingness to learn—qualities that have always defined humanity's progress. By channeling these traits, humans can shape a future where technology amplifies creativity and innovation while remaining grounded in ethical and meaningful pursuits.

10.2 Trusting the Unique Strengths of Humans and Machines

The collaboration between humans and machines marks a new era of possibilities, driven by the unique strengths each brings to the table. While humans excel in creativity, empathy, and abstract thinking, machines offer precision, speed, and data-driven insights. Together, they form a powerful partnership that has the potential to revolutionize industries, solve complex challenges, and redefine creativity. Trusting this synergy requires a deep understanding of their complementary roles and a commitment to building systems that enhance, rather than replace, human potential.

Human Creativity vs. Machine Precision

Humans and machines bring complementary strengths to the table, forming a dynamic partnership capable of driving innovation and efficiency across industries. Human creativity thrives on imagination, emotion, and the ability to synthesize disparate ideas into novel solutions. It is powered by intuition and the capacity to envision possibilities beyond the immediate or the obvious. This allows humans to craft art that stirs deep emotions, narratives that inspire action, and solutions that address both technical challenges and human needs.

Machines, on the other hand, excel in precision, speed, and scale. AI systems can analyze vast amounts of data in moments, uncovering patterns and insights that might take humans years to discover. They perform repetitive tasks with unerring accuracy and can manage immense complexity without fatigue. For example, in genomics, AI accelerates discoveries by analyzing intricate datasets, enabling breakthroughs that human effort alone could not achieve.

The synergy between human creativity and machine precision lies in their ability to complement each other. In healthcare, for example, a physician's empathetic connection with patients is augmented by AI's ability to analyze medical records, genetic data, and clinical studies in real time. Together, they form a holistic approach to diagnosis and treatment, combining human warmth with machine rigor.

In creative fields, this collaboration is equally transformative. Architects use AI to simulate designs and test structural integrity, while musicians leverage AI-generated melodies as inspiration for new compositions. Machines expand the scope of possibility, allowing humans to focus on higher-order thinking, emotional depth, and the unique touches that define originality.

Rather than seeing the relationship between humans and machines as competitive, it should be viewed as a partnership. Machines amplify human capabilities, enabling creators and professionals to achieve more than ever before. This collaboration showcases the best of both worlds—human ingenuity enriched by machine efficiency—ensuring that innovation remains impactful and deeply human.

Building Trust in Human-Machine Partnerships

Trust is the cornerstone of successful collaboration, particularly in human-machine partnerships. For such relationships to flourish, humans need assurance that machines are reliable, ethical, and aligned with their intentions, while machines must be designed to prioritize human values, adapt to user needs, and operate transparently.

Transparency as the Foundation of Trust

Building trust begins with transparency in machine operations. AI systems must be explainable and accountable, especially in high-stakes applications like healthcare, finance, and public safety. Users need clarity on how AI arrives at its conclusions or decisions. For example, an autonomous vehicle should be able to explain its actions in the event of an incident, and financial algorithms must outline the logic behind investment recommendations. Transparency demystifies technology, making it more approachable and less intimidating, which encourages users to integrate it into their workflows with confidence.

Education and Awareness: Empowering Users

Education is essential for fostering trust. When individuals understand how AI works—its capabilities, limitations, and ethical considerations—they are more likely to adopt and effectively use these tools. AI literacy programs can bridge the gap between skepticism and informed usage. For instance, teaching medical professionals how AI analyzes patient data or helping artists understand generative algorithms empowers them to collaborate with these systems without fear of obsolescence.

Educating users also dispels misconceptions, fostering a balanced perspective. This understanding reinforces the reality that while machines excel at handling complex, data-driven tasks, human skills such as empathy, intuition, and cultural interpretation remain irreplaceable.

Ethical Design: Ensuring Fairness and Accountability

Trustworthy AI begins with ethical design. Developers must prioritize fairness, inclusivity, and accountability in their algorithms. Mitigating bias, testing rigorously, and using diverse datasets are critical steps in ensuring AI operates equitably and without perpetuating harmful stereotypes. Additionally, mechanisms allowing users to question or override AI decisions ensure humans retain ultimate control.

For example, recruitment software should include safeguards to prevent discriminatory hiring recommendations, and healthcare decision-support systems should allow physicians to adjust AI-generated diagnoses based on clinical judgment.

Collaborative Environments for Trust

Creating environments where humans and machines collaborate seamlessly further strengthens trust. These settings leverage the strengths of both for optimal results. In industrial design, for instance, AI might generate rapid prototypes while engineers refine them based on intuition and practical considerations. Similarly, in disaster management, AI predicts scenarios, while human responders make strategic decisions informed by this data.

Showcasing successful human-machine collaborations can build confidence, highlighting their potential to improve efficiency, creativity, and innovation.

A Shared Future Built on Trust

Trust in human-machine partnerships is not solely about technical reliability; it is about mutual respect for the distinct strengths each brings to the table. Machines provide precision, speed, and scalability, while humans contribute emotional intelligence, intuition, and ethical judgment.

By embracing transparency, fostering education, and prioritizing ethical design, we can cultivate trust that transforms human-machine partnerships into powerful engines for progress. This trust ensures a future where technology enhances—rather than diminishes—humanity's role in shaping the world. Such a future celebrates the harmony of machine efficiency with human creativity, building a collaborative framework that drives meaningful innovation while safeguarding human values.

10.3 Coexisting with AI: Strategies for Collaboration

As AI becomes increasingly integrated into our personal and professional lives, the focus shifts from competition to collaboration. Coexisting with AI requires strategic approaches that balance human intuition and oversight with machine efficiency and precision. By defining roles, establishing trust, and leveraging complementary strengths, humans and AI can create powerful synergies.

Finding Balance: Human Oversight in AI Operations

As artificial intelligence (AI) becomes increasingly integrated into industries and daily life, effective human oversight is essential to ensure that AI aligns with ethical standards, societal norms, and user goals. Human involvement safeguards the emotional, contextual, and ethical aspects of decision-making, preserving the humanity in processes that might otherwise be dominated by automation.

Defining the Boundaries of AI Autonomy

To foster a productive human-AI partnership, clear boundaries must be established around AI's autonomy. While AI excels in analyzing data, identifying patterns, and executing tasks with speed and precision, it lacks the ability to grasp moral nuances, cultural contexts, and emotional depth. These gaps highlight the necessity of human oversight to guide AI systems toward responsible and meaningful outcomes.

Examples of Oversight in Practice

Human oversight is particularly critical in areas with significant ethical or societal implications:

✓ **Healthcare**: AI analyzes medical images, predicts patient outcomes, and identifies trends in large datasets. However, healthcare professionals must make the final decisions on diagnoses and treatments, combining AI's insights with their expertise and empathy to address patients' unique needs.

✓ **Finance**: AI algorithms optimize trading strategies, detect fraudulent activities, and analyze market trends. Yet, human judgment is indispensable for navigating economic volatility, ensuring decisions align with ethical considerations and organizational values.

✓ **Autonomous Systems**: Self-driving cars and automated factories require human oversight to manage unforeseen circumstances. Ethical dilemmas, such as prioritizing safety during accidents, demand human input to ensure decisions reflect societal priorities.

Proactive Oversight for Trustworthy AI

Oversight must be proactive rather than reactive to build trust in AI systems. Regular evaluations, audits, and updates are necessary to identify biases, inaccuracies, and unintended consequences. For example, if an AI hiring tool exhibits bias in its recommendations, oversight mechanisms can correct this before fairness is compromised. Proactive measures include:

✓ **Bias Detection**: Regularly assessing how AI processes data to prevent discriminatory outcomes.

✓ **Accuracy Verification**: Testing AI predictions against real-world scenarios to ensure continued reliability.

✓ **Accountability Tracking**: Establishing clear lines of responsibility for AI-driven decisions to maintain human accountability.

Balancing Efficiency and Accountability

While AI enhances efficiency, human oversight ensures that this efficiency does not undermine ethical or societal values. Robust governance frameworks, transparency, and interdisciplinary collaboration are vital to maintaining this balance. With active oversight, humans can guide AI to operate responsibly and effectively, leveraging its potential without compromising integrity.

Collaboration Models: Human Leadership and AI Support

Effective human-AI collaboration requires clearly defined roles, enabling a dynamic interplay where human creativity and judgment complement AI's analytical power. The collaboration model depends on the complexity, goals, and context of the task:

Human-Led with AI Support

In tasks requiring intuition, creativity, and emotional intelligence, humans lead while AI provides supportive tools:

✓ **Creative Industries**: Graphic designers use AI to generate variations, but final decisions are shaped by personal aesthetics.

✓ **Education**: Teachers use AI insights to tailor instruction but remain central to fostering intellectual and emotional growth.

✓ **Healthcare**: Physicians use AI to analyze scans, but ultimate diagnoses and treatments consider patient-specific factors.

AI-Led with Human Oversight

AI leads in data-heavy, repetitive tasks, with humans ensuring ethical and contextual alignment:

✓ **Logistics**: AI optimizes inventory and routes, while humans address cultural or environmental nuances.

✓ **Finance**: AI detects market trends, but human oversight integrates broader economic and ethical considerations.

✓ **Disaster Management**: AI predicts natural disasters, while humans strategize resource allocation based on socio-political factors.

Equal Partnership

In some scenarios, humans and AI collaborate as equals, combining their respective strengths:

✓ **Urban Planning**: AI analyzes infrastructure needs, while human planners address cultural and social dimensions.

✓ **Scientific Research**: AI processes large datasets, while researchers design experiments and interpret findings.

✓ **Content Creation**: AI generates story ideas or themes, while humans infuse emotional and cultural depth into the narrative.

Establishing Best Practices for Coexistence

To foster effective human-AI collaboration, organizations must adopt best practices that empower humans, align technology with ethical principles, and optimize the strengths of both:

✓ **Define Clear Goals**: Identify specific outcomes AI should achieve while reserving tasks requiring human intuition and empathy.

✓ **Promote Interdisciplinary Collaboration**: Combine expertise from technical, ethical, and domain-specific fields to ensure AI's responsible application.

✓ **Encourage Continuous Learning**: Equip individuals with the skills to use AI effectively while emphasizing critical thinking and ethical considerations.

✓ **Develop Ethical Frameworks**: Establish policies that address bias, transparency, and accountability, evolving them alongside AI advancements.

The Future of Human-AI Collaboration

The future of creativity and problem-solving lies in the symbiotic relationship between human ingenuity and machine intelligence. By leveraging AI's precision and scalability alongside human creativity, empathy, and judgment, individuals and organizations can address complex challenges and unlock unprecedented opportunities.

Success in this new paradigm requires cultural and organizational shifts, ensuring AI complements human abilities rather than replacing them. By adopting clear goals, fostering collaboration, and maintaining ethical integrity, human-AI partnerships can drive meaningful progress, creating a future where technology amplifies humanity's most valuable traits.

10.4 Leveraging AI for Collective Growth

The integration of AI into human workflows holds immense potential for collective growth, redefining how we work and ensuring a more equitable future. By moving beyond automation to focus on augmentation and fostering ethical collaboration, AI can empower individuals, organizations, and societies to achieve shared goals.

From Automation to Augmentation: Redefining Work

The initial impact of artificial intelligence (AI) on industries focused on automation—streamlining repetitive tasks to enhance efficiency and reduce costs. In sectors like manufacturing, logistics, and data processing, machines replicated human actions with remarkable speed and precision, revolutionizing workflows. Today, however, AI's role has evolved beyond mere automation toward augmentation, where it enhances human capabilities. This shift redefines collaboration between humans and machines, unlocking new possibilities in productivity, creativity, and innovation.

The Transition from Automation to Augmentation

Augmentation emphasizes a partnership between human expertise and machine intelligence. While AI excels in processing vast datasets, identifying patterns, and performing precise tasks, it lacks the emotional intelligence, intuition, and imagination inherent to human creativity. By integrating these complementary strengths, augmentation empowers individuals and organizations to tackle challenges previously beyond reach.

Augmentation Across Key Sectors

Healthcare: AI-powered tools in diagnostics, such as those in radiology and pathology, analyze complex data to identify conditions like cancer or heart disease with high accuracy. This enables clinicians to make quicker, more informed decisions, improving patient outcomes. However, doctors remain indispensable in interpreting AI insights, considering patient histories, and providing compassionate care. For instance, while AI may detect an anomaly in a scan, a physician evaluates its relevance within the broader context of the patient's life.

Education: In education, AI systems personalize learning by analyzing student performance, identifying areas of difficulty, and recommending tailored resources. Teachers, however, remain central in fostering critical thinking, empathy, and interpersonal skills. By integrating AI insights into lesson plans or providing targeted support, educators enhance both academic and personal development in students.

Creative Industries: Generative AI tools are transforming creative fields, assisting designers, musicians, and filmmakers in their work. A fashion designer, for instance, might use AI to forecast trends or generate design variations, refining them to produce distinctive collections. Similarly, filmmakers use AI for tasks like scene editing or lighting simulations, enabling them to focus on storytelling. In each case, AI serves as a collaborator, amplifying human creativity while leaving final decisions to the creators.

The Human-Centric Workforce

This shift ensures that humans remain at the core of the workforce, supported—not replaced—by AI. Workers in AI-augmented roles are freed from routine tasks, enabling them to focus on higher-order responsibilities requiring creativity, strategic thinking, and emotional intelligence. For example:

- ✓ **Customer Service:** AI chatbots handle routine queries, allowing human agents to address complex issues requiring empathy and critical problem-solving.

✓ **Research and Development:** AI accelerates data analysis, giving scientists more time for hypothesis generation and experimentation.

Augmentation underscores the importance of upskilling and reskilling to meet the demands of an evolving landscape. Workers must develop skills that AI cannot replicate, such as:

✓ **Strategic Thinking:** Interpreting AI-generated data for long-term planning.

✓ **Emotional Intelligence:** Applying empathy, collaboration, and communication in human-centered roles.

✓ **Creative Problem-Solving:** Innovating and adapting in response to novel challenges.

Organizations play a key role in this transition by investing in employee training, fostering lifelong learning, and providing resources to help workers effectively integrate AI into their roles.

A Collaborative Future

The transition from automation to augmentation marks a reimagining of work. AI should be seen not as a competitor but as an enabler—a tool that amplifies human strengths. This collaborative mindset fosters innovation, ensuring that technology complements human ingenuity rather than replacing it. By embracing augmentation, individuals and organizations can create environments where human potential flourishes alongside technological advancement, redefining work for a rapidly changing world.

Ethical Collaboration: Ensuring Fairness and Transparency

As artificial intelligence (AI) becomes central to decision-making across societal and organizational domains, embedding ethical principles is no longer optional—it is essential. The transformative potential of AI can only be realized if its integration prioritizes fairness, transparency, and inclusivity.

Ethical collaboration mitigates risks such as systemic bias or widening inequalities and ensures AI aligns with human values and priorities.

Mitigating Bias: Building Equitable Systems

Bias in AI remains one of the most critical ethical challenges. AI systems, trained on historical data, often risk perpetuating societal prejudices. For example:

✓ **Hiring Algorithms:** Some systems have favored male candidates due to historically biased hiring practices.

✓ **Facial Recognition:** These tools have exhibited inaccuracies across demographic groups, particularly for individuals with darker skin tones.

To combat such issues, organizations must adopt proactive measures:

✓ **Diverse Datasets:** Training AI on inclusive datasets reduces biased outcomes. For example, financial institutions can design credit-scoring models using data that encompasses diverse socioeconomic groups.

✓ **Regular Audits:** Ongoing evaluations help identify and rectify biases, ensuring systems remain fair and adapt to societal changes.

✓ **Ethical Oversight:** Multidisciplinary ethical committees, including technologists, ethicists, and community stakeholders, can assess AI deployment from a fairness perspective.

Enhancing Transparency: Building Trust

Transparency is the foundation of trust in AI. To engage stakeholders effectively, AI systems must clearly articulate how decisions are made. Explainable AI (XAI) fosters this clarity:

✓ **Explainable Algorithms:** Systems should provide comprehensible justifications for decisions. For instance, an AI rejecting a loan application could outline specific factors, such as credit history or income.

✓ **User Education:** Simplifying AI concepts through education demystifies technology and alleviates fears about "black-box" systems. This empowers users to make informed decisions and hold AI accountable.

✓ **Accountability Mechanisms:** Clear frameworks ensure humans remain responsible for AI outcomes. This includes documenting data usage, access protocols, and verifying outcomes for fairness and accuracy.

Promoting Inclusivity: Bridging the Digital Divide

Inclusivity ensures AI benefits reach all demographics, regardless of socioeconomic status, geography, or background. Without deliberate efforts, AI could deepen the digital divide. Strategies to foster inclusivity include:

✓ **Affordable Access:** Cost-effective AI tools enable adoption by smaller organizations and underserved communities.

✓ **User-Friendly Design:** Simplified interfaces and multilingual options enhance accessibility for individuals with varying technological proficiency.

✓ **Representation in Development:** Including diverse voices in AI design ensures solutions are inclusive and globally relevant.

Examples of Ethical Collaboration in Action

Ethical AI initiatives demonstrate how technology can address critical challenges while adhering to principles of fairness and transparency:

✓ **Climate Change Prediction:** AI models analyzing environmental data guide policymaking by incorporating inputs from diverse regions, benefiting both developed and developing nations.

✓ **Disaster Response:** AI systems map disaster zones and allocate resources efficiently. Transparent algorithms and fair data practices ensure aid reaches the most vulnerable.

✓ **Healthcare Equity:** AI tools for diagnosing diseases like cancer or diabetes are tailored to work effectively across populations, promoting equitable health outcomes.

Balancing Ethics and Innovation

Ethical collaboration is not about hindering innovation but about guiding it responsibly. Addressing bias, enhancing transparency, and promoting inclusivity ensures AI remains a tool for societal progress, fostering trust and confidence in its applications. By amplifying AI's capacity to create meaningful, positive change, organizations can harness its potential to build a more equitable world.

In an era increasingly shaped by technology, ethical collaboration is the key to aligning AI with humanity's collective aspirations, ensuring progress that is inclusive, transparent, and fair.

The Collective Benefits of Leveraging AI

AI holds transformative potential to address societal challenges, enhance human capabilities, and drive collective progress. By fostering fairness, transparency, and inclusivity, organizations can create a collaborative ecosystem where humans and machines work together to innovate and tackle pressing global issues.

Enhancing Productivity for Individuals and Organizations

AI excels at automating repetitive tasks and analyzing vast datasets, enabling individuals to focus on creative, strategic, and problem-solving activities. This leads to greater efficiency, cost savings, and improved decision-making across sectors:

✓ **Healthcare:** AI-powered diagnostic tools help healthcare professionals identify diseases more quickly and accurately, improving patient outcomes and enabling more personalized care.

✓ **Education:** AI enables educators to design tailored learning experiences, addressing individual student needs and driving better academic performance.

✓ **Business:** AI provides actionable customer insights, empowering businesses to deliver personalized products and services that boost customer satisfaction and foster growth.

By equipping the workforce with AI-driven tools, organizations enhance agility, innovation, and adaptability, ensuring readiness for evolving demands.

Addressing Global Challenges

AI's capacity for data analysis and predictive modeling makes it a powerful tool for tackling complex global issues:

✓ **Climate Change:** AI predicts environmental changes, optimizes renewable energy deployment, and designs sustainable urban systems, contributing to climate resilience.

✓ **Public Health:** AI enhances disease surveillance, forecasts outbreaks, and accelerates vaccine development, improving accessibility and protecting populations.

✓ **Disaster Response:** AI-driven systems enable real-time mapping of disaster zones, optimizing resource allocation to aid affected communities swiftly and effectively.

These applications demonstrate AI's ability to address challenges that require global cooperation, innovation, and long-term solutions.

Promoting Inclusivity and Reducing Inequalities

AI can democratize access to resources, narrowing gaps in education, healthcare, and economic participation. To achieve this, organizations must prioritize inclusivity in AI design:

✓ **Empowering Marginalized Communities:** Affordable and accessible AI tools enable underrepresented groups to participate in the digital economy.

✓ **Expanding Educational Opportunities:** AI-powered e-learning platforms provide quality education to remote and underserved regions.

✓ **Improving Healthcare Access:** AI-driven diagnostics and telemedicine solutions bring essential medical services to low-resource areas.

When inclusivity guides AI development, its benefits are more equitably distributed, fostering progress across diverse communities.

Transforming AI into a Catalyst for Progress

Shifting the narrative from disruption to collaboration involves ethical integration and intentional use of AI:

✓ **Fostering Public Trust:** Transparent AI practices build user confidence, encouraging adoption and collaboration.

✓ **Promoting Lifelong Learning:** Training individuals to work effectively alongside AI ensures workforce resilience and adaptability to technological advancements.

✓ **Encouraging Interdisciplinary Collaboration:** Engaging experts across fields drives innovation that is not only effective but also ethical and inclusive.

This approach ensures that AI amplifies human creativity and potential rather than diminishing it, positioning it as a catalyst for positive transformation across all facets of life.

AI as a Force for Collective Growth

By emphasizing augmentation and ethical collaboration, AI transcends its role as a tool for efficiency to become an enabler of human progress.

Thoughtfully leveraged AI fosters innovation, addresses global challenges, and creates opportunities for a more equitable and sustainable future.

Organizations that prioritize inclusivity and transparency in AI adoption can extend its benefits beyond individual gains, building a world where technology and humanity thrive in harmony. This collective growth underscores the transformative power of AI as a cornerstone of a connected, just, and prosperous society.

10.5 The Road Ahead: Preparing for a Shared Future

The integration of AI into everyday life marks a transformative moment in human history, reshaping industries, education, and the way we approach creativity and problem-solving. As we stand on the threshold of this shared future, the challenge lies in preparing individuals and institutions to embrace AI not as a threat but as a partner. This preparation requires more than technical proficiency—it demands a fundamental shift in mindset, education, and collaboration. By equipping the workforce with the skills and perspectives needed for this hybrid world, we can ensure that human ingenuity and machine intelligence work hand in hand to address complex challenges and unlock new possibilities.

Education for the Hybrid Workforce

As AI continues to reshape workplaces, the need to prepare a hybrid workforce where humans and machines collaborate seamlessly has never been more urgent. Traditional education systems, with their focus on rote memorization and theoretical learning, are no longer sufficient. The workforce of the future requires individuals who possess not just technical skills but also the creativity, adaptability, and emotional intelligence to thrive in an AI-driven world. This evolution in education must focus on equipping individuals with the tools to work alongside AI while ensuring the uniquely human qualities that machines cannot replicate remain at the forefront.

Integrating AI Literacy into Curricula

Incorporating AI literacy into education at every level is the first step toward empowering a generation that can engage meaningfully with advanced technologies. It's not enough to understand how AI works; students must also explore its applications and limitations. By combining technical knowledge with insights from the humanities, ethics, and design thinking, schools and universities can prepare individuals to navigate AI's impact on society. Interdisciplinary learning bridges the gap between technical expertise and critical thought, while practical, hands-on projects allow students to see firsthand how AI can be applied across diverse fields.

Promoting Lifelong Learning

In a world where technology evolves rapidly, learning cannot stop at graduation. Lifelong learning must become the norm, enabling individuals to update their skills as AI continues to advance. Institutions can facilitate this shift by offering accessible resources such as online courses, workshops, and certifications tailored to the changing needs of various industries. At the same time, organizations have a vital role to play by providing employees with upskilling opportunities that are aligned with workplace demands. A commitment to ongoing education ensures that individuals stay relevant and agile in an increasingly AI-integrated world.

Emphasizing Soft Skills

While AI excels at processing data and executing repetitive tasks, it is the human qualities of communication, empathy, and creativity that remain irreplaceable. Education must prioritize the development of these soft skills, ensuring that individuals complement AI's strengths with their own unique contributions. Collaborative abilities, for instance, will be critical in enabling humans to work effectively with machines, while creative thinking will allow them to harness AI as a tool for innovation. By focusing on these capabilities, educational systems can help individuals retain their relevance and impact in the workplace.

Cultivating a Growth Mindset in the Age of AI

The integration of AI into daily life and work requires not just technical skills but also a mindset that embraces change and growth. A growth mindset—the belief that abilities can be developed through effort and persistence—will be essential in adapting to the challenges and opportunities presented by AI. This mindset encourages individuals to approach uncertainties with curiosity, see setbacks as learning opportunities, and continuously push the boundaries of what is possible.

Encouraging Experimentation

The process of integrating AI often involves trial and error, making experimentation a critical part of learning. A growth mindset allows individuals to view failures not as defeats but as stepping stones toward innovation. By fostering a culture where risks are encouraged, both individuals and organizations can adapt quickly to new realities and uncover solutions that may have otherwise remained out of reach.

Redefining Success

In an AI-enhanced world, success must be redefined. It is no longer about competing with machines but about leveraging their strengths to achieve outcomes that were previously unattainable. A growth mindset helps individuals focus on collaboration, emphasizing the importance of working alongside AI to amplify human capabilities. This perspective not only alleviates fears of obsolescence but also encourages continuous improvement and learning.

Fostering Resilience

Rapid advancements in AI can create uncertainties, particularly in industries undergoing significant transformation. Resilience—the ability to adapt and recover—becomes a crucial trait for navigating such shifts. With a growth mindset, individuals are better equipped to face these changes head-on, viewing challenges as opportunities for reinvention and personal growth.

Promoting Inclusive Thinking

A growth mindset is not just about personal development; it also encourages inclusivity. Recognizing the value of diverse perspectives fosters collaboration that transcends cultural, disciplinary, and experiential boundaries. In an AI-driven world, this inclusivity fuels innovation by bringing together varied viewpoints to solve complex problems in creative ways.

Paving the Way for a Collaborative Future

The future of work is no longer defined by human effort alone but by the symbiotic relationship between humans and AI. Preparing for this collaborative future requires more than just technical training; it demands the cultivation of adaptability, curiosity, and a lifelong commitment to learning. By embedding AI literacy into education, promoting soft skills, and fostering a growth mindset, we can equip individuals to navigate the complexities of an AI-augmented world.

This forward-thinking approach ensures that humanity remains central to progress. It shifts the narrative from machines replacing humans to machines enhancing human potential. With the right preparation, the future holds extraordinary possibilities, where humans and AI work together to create innovations that not only advance industries but also enrich lives in meaningful ways.

Summary

This chapter envisions a future where human creativity and machine intelligence coexist harmoniously, forming a partnership that amplifies human potential. It explores how collaboration between humans and machines can drive innovation, creativity, and problem-solving while emphasizing the importance of maintaining human oversight, ethical responsibility, and intentionality in the process.

Key Themes:

A Synergistic Relationship: The chapter highlights the complementary strengths of humans and machines: humans bring emotional intelligence, imagination, and cultural context, while machines contribute speed, precision, and data analysis.

It frames this relationship as an opportunity to transcend the limitations of both parties, unlocking new creative and practical possibilities.

Augmenting Creativity: Machines are seen as enablers, not replacements. They enhance the creative process by handling repetitive tasks, generating ideas, and offering unique insights from data.

Examples include filmmakers using AI to storyboard complex scenes or architects leveraging AI to simulate environmental impact.

The Role of Human Oversight: While machines can perform tasks efficiently, human creators must retain control over critical decisions to ensure authenticity and relevance.

The chapter emphasizes the need for intentionality, where humans guide machines to align with ethical, emotional, and cultural values.

Ethical Collaboration: The chapter addresses the ethical implications of integrating machines into creative workflows, urging creators to ensure fairness, inclusivity, and accountability.

Transparency in acknowledging AI's role in the creative process builds trust and ensures the final output reflects human values.

Learning from Machines: Machines can inspire humans by uncovering patterns or solutions that may not be immediately apparent, pushing creators to explore uncharted territories.

However, humans must approach AI-generated suggestions critically, evaluating their context, quality, and potential impact.

The Evolution of Roles: The chapter predicts a shift in roles where machines handle analytical and repetitive tasks, freeing humans to focus on innovation, strategic thinking, and emotional depth.

For example, in education, AI might personalize learning paths, while teachers focus on fostering critical thinking and interpersonal skills.

Collaboration Across Disciplines: True innovation emerges when AI-powered tools bring together experts from diverse fields. For instance, AI can facilitate projects that require the intersection of engineering, art, and sociology.

Such collaborations leverage both machine-driven insights and human creativity to address complex societal challenges.

Preparing for the Collaborative Future: The chapter calls for a mindset shift, where individuals see AI as a partner rather than a competitor.

Lifelong learning, ethical awareness, and adaptability are identified as essential traits for thriving in this collaborative future.

Balancing the Human and Machine Contributions: The chapter advocates for maintaining a balance where machines handle efficiency and scalability, but humans shape creativity, empathy, and ethical considerations.

This balance ensures that technological advancements amplify human ingenuity rather than diminish it.

The collaborative future envisioned in this chapter emphasizes the potential for humans and machines to work together in a way that enhances creativity, innovation, and societal progress. By cultivating a thoughtful approach to this partnership—grounded in ethics, intentionality, and adaptability—individuals and institutions can ensure that the integration of AI serves to complement humanity's strengths rather than replace them. This collaboration holds the promise of a transformative era, where technology and humanity work together to achieve extraordinary outcomes.

Part V: Solutions and Actionable Frameworks

"Creativity thrives when guided by vision and action—where solutions are not just imagined but implemented, turning challenges into opportunities and ideas into impactful realities."

— Sandeep Chavan

11: Rethinking Policies

As the influence of artificial intelligence continues to reshape industries and societies, educational and research institutions find themselves at a critical juncture. The policies governing these institutions must evolve to balance ethical considerations with the need for innovation, ensuring that AI is integrated thoughtfully into academic environments. While tradition has long been the foundation of institutional practices, the rapid pace of technological advancement demands a forward-thinking approach that bridges the gap between the past and the future. This chapter explores strategies for ethical AI integration, fostering collaboration, and creating adaptive policies that empower institutions to lead in an era defined by intelligence—both human and artificial.

11.1 The Need for a Policy Overhaul in the AI Era

In an age where artificial intelligence is redefining how we learn, work, and innovate, institutional policies must evolve to keep pace with this transformative force. Traditional frameworks, while rooted in valuable principles, often struggle to address the complexities of AI's integration into education and research. The rapid advancement of AI demands a fresh perspective—one that balances ethical considerations with the urgency of staying relevant in a competitive, tech-driven world.

Institutions are not only the custodians of knowledge but also the architects of the future. Their policies shape how students, researchers, and society at large engage with emerging technologies. Yet, overly cautious approaches can stifle creativity, hinder collaboration, and delay the adoption of tools that have the potential to revolutionize learning and discovery.

A policy overhaul is not merely about adapting to change; it's about proactively shaping the future. Institutions must move beyond rigid rules to embrace adaptive, forward-thinking policies that enable experimentation while safeguarding ethical values. By reimagining their role in this era of rapid innovation, they can empower individuals and communities to harness AI's potential responsibly and effectively.

11.2 Ethical Integration of AI in Education and Research

As AI becomes integral to education and research, institutions must ensure its integration adheres to ethical principles. This responsibility goes beyond imparting technical skills, demanding a commitment to transparency, accountability, and a values-driven approach in academic practices. By embedding ethical considerations into policies, curricula, and research methodologies, universities and colleges can ensure AI becomes a force for equitable progress and innovation.

Embedding AI Ethics in Curricula

Ethics should serve as the foundation of any AI education program. While technical proficiency equips students to build AI systems, a robust understanding of ethical challenges ensures these tools are applied responsibly. Core topics such as algorithmic bias, data privacy, accountability, and societal impact must be seamlessly integrated into AI courses, preparing students to navigate the complexities of a rapidly evolving technological landscape.

Interdisciplinary approaches enrich this learning. Combining computer science with disciplines like philosophy, sociology, and legal studies provides students with a comprehensive understanding of AI's broader implications. This fosters critical thinking, enabling future innovators to evaluate the societal consequences of their creations. For example, students could analyze case studies on biased algorithms or debate the ethics of surveillance technologies, grounding their technical expertise in accountability and social responsibility.

Ensuring Transparency in AI-Driven Academic Tools

AI tools are becoming indispensable in academia, supporting tasks such as automated grading, plagiarism detection, and personalized learning. While these tools enhance efficiency and enable educators to focus on nuanced aspects of teaching, their opaque nature can raise concerns about fairness and trust. To foster confidence, it is vital that students and educators understand how these systems operate and make decisions.

Explainable AI (XAI) can address these concerns by providing clear insights into algorithmic processes. For instance, if an AI system assigns grades, it should be able to explain the criteria and factors influencing its decisions. Similarly, personalized learning platforms should transparently outline the logic behind their recommendations, empowering students to make informed choices about their education. Regular audits and open communication regarding the capabilities and limitations of AI tools are also essential. These practices demystify AI, positioning it as a partner in education rather than an opaque authority.

Encouraging Responsible Use of AI in Research

AI offers unprecedented opportunities in research, from analyzing massive datasets to simulating complex scenarios. However, its power demands a heightened sense of responsibility to ensure alignment with ethical standards and societal goals. Institutions must establish clear guidelines for responsible AI research, prioritizing transparency, inclusivity, and integrity.

Transparency is particularly critical in research methodologies. Researchers should meticulously document processes, datasets, and algorithms, ensuring their work can be replicated and scrutinized. This not only upholds academic integrity but also builds trust in AI-driven discoveries. Furthermore, institutions should advocate for the use of diverse and representative datasets to prevent perpetuating biases in research outcomes. For example, healthcare AI projects must include demographic diversity to ensure equitable treatment recommendations.

Interdisciplinary collaboration enriches the ethical application of AI in research. A climate modeling project, for instance, can benefit from the perspectives of environmental scientists, data analysts, and social scientists to holistically address its implications. Institutions should support such initiatives by establishing ethics committees or review boards to evaluate AI-related projects, ensuring alignment with societal values and addressing global challenges.

Leading the Way in Ethical AI Integration

By embedding AI ethics into curricula, ensuring transparency in academic tools, and promoting responsible research practices, institutions can establish a benchmark for ethical AI integration. This approach prepares students and researchers to tackle the challenges of the AI era with integrity, creativity, and accountability. It also reaffirms education's critical role as a guardian of human values amid rapid technological transformation, ensuring that progress benefits humanity equitably and responsibly.

In championing these principles, academic institutions not only equip future innovators with the skills to leverage AI but also cultivate a generation committed to using technology for the greater good.

11.3 Bridging the Gap Between Tradition and Innovation

In a time marked by rapid technological advancements, educational institutions face the challenge of harmonizing progress with their enduring traditions. The task is not just to adopt emerging technologies but to thoughtfully integrate them with academic principles that have shaped intellectual development for centuries. By blending innovation with heritage, universities can preserve the values of mentorship, rigorous inquiry, and intellectual growth while equipping students to thrive in an increasingly complex and dynamic global landscape.

Preserving Academic Heritage While Embracing Technology

The academic heritage of universities—rooted in critical thinking, debate, and the pursuit of knowledge—continues to be a cornerstone of intellectual growth. These traditions cultivate curiosity, deepen understanding, and foster rigorous inquiry. Yet, the transformative potential of technology can enhance these timeless principles, offering new pathways for exploration and learning.

For instance, the Socratic method, which encourages dialogue and critical questioning, can be revitalized with AI-powered tools that provide real-time data, diverse perspectives, and simulated scenarios to enrich discussions. Similarly, digitization has revolutionized archival research, making rare historical materials accessible to students and researchers worldwide. This interplay between tradition and technology preserves the depth and rigor of academic practices while equipping students with modern tools to navigate a technology-driven world.

Such a harmonious integration ensures that innovation amplifies the essence of academic traditions rather than overshadowing them, fostering a balance that respects the past while embracing the future.

Updating Pedagogical Approaches for a Hybrid World

The modern classroom has evolved into a hybrid learning environment, blending physical and digital spaces to meet the needs of today's learners. This shift requires rethinking traditional teaching methods to engage students who are accustomed to interactive and adaptive learning experiences.

AI-powered platforms hold the potential to revolutionize pedagogy by analyzing individual learning styles and tailoring content to diverse needs. For example, AI systems can identify students' strengths and weaknesses, offering personalized exercises to enhance their understanding. Meanwhile, technologies like virtual reality (VR) and augmented reality (AR) enable immersive educational experiences—a history lesson might transport students to ancient civilizations, or a biology class could use AR to visualize cellular processes, turning abstract concepts into tangible experiences.

However, adopting technology is only part of the equation. Equally important is fostering digital literacy alongside critical thinking, ensuring students not only master modern tools but also understand their societal and ethical implications. This dual approach prepares learners to contribute meaningfully to a world where technology and humanity coexist, blending proficiency with thoughtful analysis.

Promoting Cross-Disciplinary Collaboration

Innovation often arises at the intersections of diverse disciplines, where different perspectives converge to address complex challenges. Universities are uniquely positioned to foster such cross-disciplinary collaboration, equipping students with the tools to think beyond the confines of their chosen fields.

For example, using AI in healthcare diagnostics can benefit from the combined expertise of computer scientists, medical professionals, ethicists, and sociologists. Such interdisciplinary efforts not only yield robust solutions but also ensure they are equitable and socially responsible. Similarly, partnerships between artists and engineers can drive groundbreaking innovations in design, media, and technology, blending creativity with technical precision.

To encourage these synergies, universities must establish interdisciplinary programs, create research hubs that foster collaboration, and provide platforms for faculty and students to engage across fields. These initiatives cultivate adaptability, teamwork, and problem-solving skills, empowering students to tackle multifaceted challenges in an interconnected world.

Building a Bridge for the Future

Bridging the gap between tradition and innovation is not about choosing one over the other—it is about integrating both. By preserving the core values of academia, updating pedagogical methods to reflect hybrid realities, and promoting cross-disciplinary collaboration, universities can create a dynamic educational environment that prepares students for the demands of the future.

This balanced approach ensures that students emerge not only as skilled professionals but as thoughtful, creative, and socially responsible citizens. When education harmonizes tradition with innovation, it becomes a transformative force that nurtures leaders equipped to navigate and shape a rapidly evolving world. In doing so, institutions can uphold the enduring relevance of education while positioning it as a beacon of progress and humanity in the digital age.

11.4 Building Collaborative Ecosystems for Institutions

In the age of AI, educational institutions face unparalleled challenges and opportunities that demand collaboration on a global scale. No single university, organization, or nation can address the complexities of an AI-driven future in isolation. To remain relevant and impactful, institutions must build collaborative ecosystems that bridge academia, industry, and diverse global stakeholders. These ecosystems will not only accelerate innovation and learning but also ensure that AI evolves in an ethically responsible manner. By advancing knowledge collectively and addressing societal challenges, such partnerships will shape a future where AI benefits all.

Strengthening Partnerships Between Academia and Industry

The disconnect between theoretical research and practical application has long been a challenge in education. Strengthening ties between academia and industry can bridge this gap, yielding mutual benefits. Universities gain access to real-world problems, funding, and insights, while industries tap into fresh talent and innovative ideas.

For example, industry-led projects integrated into academic programs allow students to work on tangible challenges, such as developing AI solutions for energy efficiency or public health crises. These initiatives foster practical experience while addressing critical global issues. Joint research programs can focus on emerging fields like renewable energy or sustainable agriculture, creating innovations with far-reaching societal impact.

Industry-sponsored research labs within academic settings can serve as incubators for cutting-edge technologies. These labs facilitate collaboration between faculty, students, and professionals, fostering a two-way exchange of knowledge and expertise. Such partnerships also align academic curricula with market demands, ensuring graduates are equipped with relevant skills.

Lifelong learning programs facilitated by universities further strengthen academia-industry ties. These programs enable professionals to return to academia, updating their skills to keep pace with technological advancements. This cyclical relationship between learning and application promotes continuous growth and innovation.

Creating Open Platforms for AI Innovation and Learning

Accessibility and inclusivity are crucial for advancing AI on a global scale. Open platforms for research and education democratize access to AI tools, datasets, and resources, empowering individuals and institutions across socio-economic and geographical boundaries.

Universities can establish open-source repositories where researchers, students, and industry professionals collaborate to develop AI models, share datasets, and test solutions. For example, an open-access platform focused on AI for climate change could invite contributions from experts worldwide, fostering collective progress while maintaining transparency.

In education, open platforms can revolutionize learning. Virtual labs, interactive tutorials, and open-access courses enable learners from diverse backgrounds to gain practical AI experience. AI-powered simulations can immerse students in real-world scenarios, such as managing disaster responses or analyzing financial markets, offering hands-on learning opportunities that bridge theory and application.

By breaking down barriers to entry, these platforms ensure that the benefits of AI education and innovation are accessible to a broader audience, nurturing talent and expanding participation in AI's evolution.

Fostering Global Collaboration in Ethical AI Development

The global implications of AI necessitate collaboration across borders, cultures, and disciplines. Ethical AI development demands diverse perspectives to address universal challenges while respecting cultural and regional nuances. Universities, as hubs of intellectual exchange, play a critical role in fostering such international partnerships.

Global summits, workshops, and joint research initiatives can bring together academia, government, and industry stakeholders to share best practices and align regulatory frameworks. For instance, an international consortium on ethical AI could focus on establishing standards for transparency, fairness, and accountability, ensuring equitable benefits across populations.

Collaborative projects that address global challenges, such as food security or disaster mitigation, highlight the power of cross-border efforts. For example, an AI model co-developed by institutions in developed and developing countries could combine advanced analytics with local expertise, yielding solutions that are innovative yet contextually relevant.

Inclusion is essential for effective global collaboration. Actively involving underrepresented communities and regions ensures their perspectives shape AI development. Incorporating indigenous knowledge into environmental AI models, for instance, can lead to more sustainable and inclusive outcomes that honor both scientific and cultural wisdom.

A Unified Ecosystem for Progress

Collaborative ecosystems are not just an option—they are imperative for leveraging AI responsibly and effectively. Stronger ties between academia and industry ensure that innovation is both practical and impactful. Open platforms democratize access to tools and resources, enabling widespread participation in AI's growth. Global partnerships foster diverse perspectives, creating solutions that are equitable, culturally sensitive, and transformative.

By building such ecosystems, educational institutions can lead the charge in shaping an AI-driven future grounded in ethical responsibility and societal progress. These collaborations position AI not merely as a tool of transformation but as a unifying force, enabling shared growth and development for individuals, societies, and the planet. In doing so, institutions underscore AI's potential to advance humanity collectively, creating a more inclusive, equitable, and sustainable world.

11.5 Implementing Adaptive Policies for Future Readiness

In an era of rapid technological advancements, educational institutions must adopt adaptive policies to remain relevant and impactful. These policies should bridge the gap between innovation and regulation, preparing institutions to address the challenges and opportunities of an AI-driven world. By fostering flexibility, promoting lifelong learning, and establishing robust ethical frameworks, institutions can enhance their effectiveness and amplify their societal impact in the AI age.

Balancing Regulation with Flexibility

Adaptive policies require striking a balance between fostering innovation and maintaining ethical standards. Overregulation can stifle creativity and delay progress, while insufficient oversight risks ethical violations, inequities, and unforeseen societal consequences. Institutions must create frameworks that evolve with technological advancements, promoting responsible experimentation while upholding accountability.

Regulatory sandboxes offer an effective solution, providing controlled environments for testing AI systems. Universities can experiment with new technologies in these spaces, assessing their safety, ethical implications, and practical benefits before full-scale implementation. Similarly, **data usage policies** should integrate techniques like anonymization and secure sharing to enable researchers to access critical datasets while protecting privacy and ensuring compliance with data protection laws.

Flexibility in policies empowers researchers and educators to explore AI's transformative potential responsibly. This balanced approach fosters innovation without compromising ethical or societal values.

Supporting Lifelong Learning and Faculty Development

Continuous learning is vital for equipping students, educators, and researchers with the skills and knowledge needed to thrive in a rapidly evolving technological landscape. Institutions must invest in initiatives that ensure their communities remain adaptable and future-ready.

For Faculty: Professional development programs focused on AI and related fields are essential for keeping educators at the forefront of technological advancements. Workshops, certifications, and sabbaticals can help faculty integrate cutting-edge tools into their teaching while addressing AI's ethical dimensions. For instance, a literature professor might use AI for textual analysis, while a biology instructor employs it to model ecosystems. These initiatives ensure educators remain technologically adept while contextualizing AI within their disciplines.

For Students: Lifelong learning opportunities should extend beyond traditional degree programs to include modular courses, micro-credentials, and industry partnerships. Real-world learning experiences, such as internships, mentorships, and cooperative programs, help students bridge the gap between academia and the workforce. By preparing students for the demands of a hybrid world, institutions ensure their graduates remain competitive and adaptable in an AI-driven society.

Establishing Metrics for Ethical and Impactful AI Integration

Clear, measurable criteria are essential for evaluating the ethical and practical impacts of AI integration. These metrics guide the implementation and assessment of AI systems, ensuring they align with educational and societal goals.

Transparency and Fairness: Metrics for academic AI tools, such as automated grading systems, should prioritize transparency, fairness, and user satisfaction. Students should understand how grades are determined, and educators must be able to audit and refine these systems to maintain equity.

Societal Relevance: Research-driven AI initiatives should address global challenges, from climate change to healthcare inequities. Metrics should evaluate the inclusivity and societal impact of these projects, emphasizing work that benefits underserved communities.

Audits and Accountability: Regular audits can identify biases, inaccuracies, or unintended consequences in AI systems. Interdisciplinary audit teams, comprising technologists, ethicists, and domain experts, ensure that AI tools align with institutional values and objectives.

Collaborative Feedback: Institutions should implement feedback loops that engage students, faculty, and external stakeholders in evaluating AI initiatives. This inclusive approach fosters trust, responsiveness, and continuous improvement, ensuring AI systems evolve to meet the needs of all stakeholders.

Shaping the Future Through Adaptation

Adaptive policies are the foundation of future-ready institutions. By balancing regulation with flexibility, supporting lifelong learning, and establishing robust ethical metrics, universities can effectively navigate the complexities of the AI era. These policies empower institutions to embrace technological innovation while safeguarding their commitment to equity, excellence, and societal well-being.

Through thoughtful adaptation, educational institutions can lead the way in shaping a collaborative, inclusive, and transformative future. By preparing students, faculty, and society to thrive in a world where AI plays an integral role, these policies ensure that progress serves humanity equitably and responsibly.

11.6 Shaping Institutions as Catalysts for Societal Progress

In the age of AI, educational institutions are uniquely positioned to go beyond their traditional roles and act as catalysts for societal transformation. By leveraging AI for community development, fostering equity in education, and establishing themselves as leaders on the global stage, universities can tackle pressing social challenges while championing ethical innovation and inclusivity.

Leveraging AI for Community and Social Development

Educational institutions have the resources and expertise to harness AI in ways that directly benefit local and global communities. By applying AI research to real-world problems, universities can create solutions that extend far beyond academic settings.

For example, **AI-driven public health initiatives** can enhance early disease detection, track outbreaks, and deliver personalized care in underserved regions, addressing critical healthcare disparities. In **environmental sustainability**, institutions can lead projects to model climate change impacts, optimize resource utilization, and innovate green technologies. Collaborating with local governments, universities can also apply AI to improve urban planning, streamline traffic management, and bolster disaster response, creating safer and more resilient cities.

By aligning their AI capabilities with community needs, institutions not only contribute to societal well-being but also position themselves as indispensable partners in addressing global challenges.

Promoting Equity and Inclusivity in AI Education

To ensure that the benefits of AI are widely shared, institutions must prioritize equity and inclusivity in AI education. Marginalized communities and underrepresented groups must have equal opportunities to participate in and benefit from the AI revolution.

Strategies for Equity

✓ **Accessible Programs:** Universities can offer low-cost or free AI education initiatives, such as online courses, workshops, and certifications, to reach underserved populations. Scholarships and grants for students from underrepresented groups can further bridge gaps in AI expertise.

✓ **Community Outreach:** Partnering with schools and local organizations to introduce AI concepts at an early age can inspire interest in AI-related careers. Hands-on camps, mentorship programs, and outreach in areas with limited technological access can spark curiosity and build foundational knowledge.

✓ **Inclusive Curriculum:** AI curricula should celebrate diverse contributions from different cultures, genders, and regions. Highlighting a range of perspectives helps dismantle stereotypes, fosters a sense of belonging, and encourages participation from historically excluded groups.

By promoting inclusivity, institutions can democratize AI education and empower a broader range of individuals to influence the future of technology.

Preparing Institutions for Long-Term Global Leadership

To remain influential in a rapidly evolving global landscape, universities must adopt strategies that prioritize collaboration, adaptability, and societal impact, positioning themselves as pioneers of ethical AI innovation.

✓ **Building Global Networks:** Active participation in international AI initiatives enables institutions to collaborate with universities, industries, and policymakers worldwide. By contributing to global research projects—such as AI for climate resilience or pandemic preparedness—universities can demonstrate their leadership in addressing shared challenges.

✓ **Investing in Talent:** Attracting and retaining top talent is vital for maintaining a competitive edge. Universities can achieve this by fostering environments that encourage groundbreaking research, provide access to cutting-edge resources, and champion diversity in both faculty and student recruitment. Dedicated fellowships and postdoctoral programs focusing on AI ethics and applications can draw the brightest minds to their institutions.

✓ **Shaping Policy and Thought Leadership:** Universities must play a leading role in shaping global AI policies by offering evidence-based recommendations and engaging with governments and international bodies. Thought leadership in ethical AI governance, inclusivity, and sustainable development not only enhances institutional reputation but also ensures a seat at the table in critical policy discussions.

A Vision for Societal Impact

By leveraging AI for community development, championing equity in education, and preparing for global leadership, educational institutions can redefine themselves as engines of societal progress. These efforts ensure that AI evolves as more than a tool for technological advancement—it becomes a force for building a more equitable, inclusive, and sustainable world.

Through strategic initiatives and an unwavering commitment to ethical innovation, universities have the power to inspire change, empower communities, and lead humanity toward a brighter, more interconnected future.

Summary

This chapter focuses on the critical need for educational institutions to reform their policies to adapt to the AI-driven era. It offers a roadmap for creating future-ready frameworks that align education with technological advancements, promote ethical innovation, and empower students, educators, and researchers. The chapter emphasizes balancing tradition with innovation while ensuring inclusivity, adaptability, and societal relevance.

Key Themes:

Adapting Policies for the AI Age: Institutions must implement adaptive policies to keep pace with rapid technological advancements. These policies should balance regulation and flexibility, allowing institutions to foster innovation while maintaining ethical oversight. Regulatory sandboxes are introduced as a mechanism to experiment with AI technologies in controlled environments, ensuring safe and impactful implementation.

Promoting Lifelong Learning: Lifelong learning is identified as essential for both students and faculty to remain relevant in a rapidly evolving world. Faculty development programs should include training on AI tools and their ethical implications to enhance teaching and research. Modular courses, certifications, and partnerships with industry can help students and professionals continuously upgrade their skills.

Ethical and Impact-Driven Integration of AI: Institutions must establish clear metrics to evaluate the ethical and societal impact of AI integration in education and research. Transparency, fairness, and inclusivity should be foundational principles guiding the use of AI.

Audits and interdisciplinary review boards can ensure that AI tools align with institutional values and societal needs.

Bridging Academia and Industry: Strengthening partnerships with industry is crucial for bridging the gap between theoretical research and practical applications. Collaborative projects, internships, and industry-sponsored research labs can align academic efforts with real-world challenges. These collaborations prepare students for the workforce and contribute to solving pressing global issues like climate change and healthcare disparities.

Fostering Interdisciplinary Collaboration: Innovation often emerges at the intersection of disciplines. Institutions should encourage collaboration across fields to tackle complex challenges. Research centers and interdisciplinary programs can provide platforms for students and faculty to work on projects that combine technology with arts, humanities, and social sciences.

Ensuring Inclusivity in AI Education: Institutions must address the digital divide by making AI education accessible to underserved communities. Low-cost or free courses, scholarships, and community outreach programs can democratize access to AI tools and knowledge. Inclusive curricula that reflect diverse cultural perspectives ensure that the benefits of AI are distributed equitably.

Global Leadership in Ethical AI Development: Universities have a role in shaping global AI policies and promoting thought leadership in ethical AI practices. Participation in international research initiatives and policy discussions allows institutions to influence the global AI landscape positively. Building networks with global stakeholders ensures that AI solutions address universal challenges while respecting cultural nuances.

Balancing Tradition and Innovation: While embracing technological advancements, institutions must preserve the intellectual and ethical heritage that defines academia. Policies should encourage the integration of AI tools without compromising the depth, rigor, and cultural relevance of traditional educational practices.

EXTENSION: FROM BRUSH TO ALGORITHM

This chapter calls for a holistic rethinking of institutional policies to ensure that education remains impactful and relevant in the AI era. By fostering adaptive frameworks, promoting ethical practices, and prioritizing inclusivity, institutions can position themselves as leaders in shaping a collaborative, innovative, and equitable future. These reimagined policies not only empower students and educators but also align academic efforts with societal progress, ensuring that technology serves humanity responsibly and effectively.

12: Balanced Creative Identity

The rise of AI has redefined the boundaries of creativity, offering creators unprecedented tools to enhance their work while challenging traditional notions of originality and authorship. To thrive in this new era, creators must embrace the opportunities provided by AI while preserving their unique human touch. Empowering creators in the machine age requires a focus on building AI literacy and fostering a balanced creative identity that integrates human ingenuity with technological innovation.

12.1 Building AI Literacy

AI literacy has become a cornerstone of creative empowerment in the machine age. It is no longer an optional skill but a critical necessity for creators aiming to thrive in a technology-driven world. Understanding AI's capabilities, limitations, and ethical implications equips creators to harness it effectively and responsibly. Rather than posing a threat, AI serves as a collaborator—an enabler of human ingenuity and innovation. Developing AI literacy ensures that creators can seamlessly integrate AI into their workflows while preserving their unique voices and intentions.

Understanding the Basics

AI literacy begins with a foundational grasp of key concepts such as machine learning, neural networks, data processing, and generative algorithms. While creators don't need to delve deeply into technical intricacies, having a working knowledge of these ideas empowers them to interact with AI tools meaningfully and maximize their potential.

For example, a photographer using AI for image enhancement benefits from understanding how algorithms process light, texture, and color to refine visuals. Similarly, a writer collaborating with a language model can achieve better results by knowing how the AI interprets context and predicts word patterns. This foundational knowledge enables creators to approach AI with clarity, tailoring its outputs to align with their creative goals.

Accessible resources, such as online courses, tutorials, and community forums, offer valuable opportunities to build this understanding. By demystifying complex concepts, these resources make AI tools approachable and practical for professionals across diverse industries.

Hands-On Experience

Practical experience is essential for building AI literacy. Direct engagement with AI tools allows creators to experiment, learn through trial and error, and discover how to effectively incorporate AI into their workflows.

Platforms like ChatGPT, DALL-E, and MidJourney, as well as workshops and hackathons, provide hands-on opportunities for creators to explore AI's potential. For instance, a visual artist might use AI to generate preliminary sketches and refine them into polished works, while a musician could draw inspiration from AI-generated melodies to craft original compositions. Such interactions reveal the vast possibilities of AI while also highlighting its limitations and areas for improvement.

By bridging the gap between conceptual knowledge and real-world application, practical exposure helps creators build the confidence and adaptability needed to innovate without compromising their artistic vision.

Critical Thinking and Ethical Awareness

AI literacy extends beyond technical skills; it encompasses critical thinking and a strong ethical foundation. Creators must recognize that AI outputs are shaped by the data they are trained on, which can introduce biases or reinforce stereotypes. Developing this awareness enables creators to use AI responsibly, ensuring their work reflects fairness, inclusivity, and integrity.

For example, a filmmaker using AI to generate dialogue should critically evaluate whether the suggestions perpetuate cultural or gender biases. Similarly, a product designer employing AI for user interface development must ensure accessibility for diverse populations. Scrutinizing AI-generated outputs allows creators to mitigate potential harm and align their work with ethical standards.

Transparency is another vital aspect of ethical AI use. Acknowledging AI's role in the creative process fosters trust and promotes openness. For instance, a writer leveraging AI for brainstorming might share how the tool influenced their ideas, highlighting the collaboration between human intuition and machine capability.

The Benefits of AI Literacy

Investing in AI literacy equips creators with the knowledge and confidence to navigate the complexities of the machine age. It transforms AI from a perceived threat into a powerful ally, enhancing productivity, expanding creative horizons, and opening new avenues for innovation.

By combining foundational understanding, hands-on experience, and ethical awareness, creators can harness AI's potential while safeguarding the authenticity and intent that define their work. This approach ensures that creativity remains a distinctly human endeavor, enriched and amplified by the tools of the modern age.

12.2 Developing a Balanced Creative Identity

In the machine age, creators face the challenge of integrating AI's capabilities into their work without compromising the authenticity that defines human creativity. Striking this balance ensures that their creations remain deeply personal, culturally resonant, and innovative. A balanced creative identity allows creators to harness the strengths of AI while preserving their unique voice, purpose, and connection to their craft.

Preserving Human Authenticity

Authenticity is at the core of every meaningful creative endeavor. Creativity draws from human emotion, intuition, and lived experience—qualities AI cannot replicate. While AI can assist by generating ideas or executing technical tasks, the essence of creative work must originate from the creator's personal vision and voice.

For example, a writer might use AI to suggest character arcs or plot twists but will infuse the narrative with their emotional depth, cultural nuances, and individual perspective. Similarly, a visual artist experimenting with AI-generated patterns ensures that the final piece reflects their interpretation and intention. This approach positions AI as a complementary tool, enhancing creativity without replacing the creator's authenticity.

Combining Strengths

A balanced creative identity thrives on the synergy between human imagination and machine precision. AI excels at analyzing data, recognizing patterns, and iterating rapidly, while humans bring empathy, contextual understanding, and the ability to envision entirely new possibilities. Together, these strengths unlock extraordinary creative breakthroughs.

For instance:

✓ **In Music:** AI can compose harmonies or melodies, but the human musician adds emotional tone, lyrics, and instrumentation that deeply connect with audiences.

✓ **In Architecture:** AI might generate structural designs based on environmental data, but the architect integrates cultural and aesthetic elements to create spaces that feel alive and meaningful.

By blending AI's analytical power with human emotional and imaginative depth, creators can transcend traditional limitations and push the boundaries of possibility.

Adaptability and Growth

In a rapidly evolving technological landscape, adaptability and a growth mindset are essential for creators. AI should not be viewed as a static tool but as a dynamic collaborator that evolves alongside their craft.

Adaptability means experimenting with new workflows, embracing emerging tools, and reimagining traditional approaches. For example, a filmmaker might explore AI-generated storyboards or visual effects to enhance their storytelling while maintaining control over the narrative's emotional core. Continuous learning—through workshops, online courses, or self-directed exploration—ensures that creators remain agile and confident in navigating a machine-enhanced world.

This mindset fosters resilience and encourages creators to push boundaries, innovate, and stay grounded in their unique identity.

Balancing Innovation with Ethics

A balanced creative identity must also address the ethical implications of AI usage. While AI amplifies creative potential, it can also introduce risks, such as copyright infringement, bias, or unintended consequences. Creators must use AI thoughtfully, ensuring their innovations align with ethical standards and positively impact society.

For example:

✓ **In Visual Arts:** Artists should ensure AI-generated works do not replicate existing intellectual property or reinforce stereotypes.

✓ **In Writing:** Authors using AI for drafting must ensure outputs reflect inclusive storytelling principles and avoid unintentional bias.

Ethical considerations guide creators in respecting intellectual property, cultural diversity, and societal values. Transparency about AI's role in the creative process builds trust with audiences, reinforcing the integrity of the creator's vision.

Redefining Creativity in the Machine Age

Developing a balanced creative identity is not about choosing between human creativity and machine assistance—it is about harmonizing the two. By preserving authenticity, combining strengths, fostering adaptability, and upholding ethical standards, creators can redefine creativity for the machine age. This balanced approach empowers creators to embrace technology as a partner in innovation while staying true to the emotional depth and unique perspectives that make human creativity timeless.

12.3 A Future of Empowered Creators

The machine age offers creators an unparalleled opportunity to elevate their craft by blending human ingenuity with technological innovation. Empowering creators in this era is not about replacing their unique contributions with technology but about augmenting their capabilities. By embracing AI literacy and cultivating a balanced creative identity, creators can harness the tools of the machine age while preserving the emotional depth, authenticity, and originality that define their work.

Embracing Collaboration for Innovation

The future of creativity lies in collaboration—between humans and machines, tradition and innovation, and individual expression and collective progress. This synergy unlocks new dimensions of creative potential, where AI provides efficiency, precision, and computational power to complement the intuition, imagination, and empathy of human creators.

For example:

✓ **In Storytelling:** AI can suggest narrative structures or character archetypes, while the writer weaves these elements into a meaningful story enriched by personal experience.

✓ **In Design:** AI can generate numerous variations of patterns or forms, enabling the creator to select, refine, and infuse personal touches into the final product.

✓ **In Scientific Discovery:** AI accelerates data analysis, but human researchers interpret findings, draw connections, and explore hypotheses that push the boundaries of knowledge.

This collaborative model ensures that technology enhances, rather than overshadows, human creativity, enabling creators to focus on the aspects of their work that inspire connection and meaning.

Preserving the Essence of Creativity

Amid rapid technological advancement, the essence of creativity—expression, storytelling, and innovation—must remain firmly rooted in human intent. Empowered creators use AI tools not as substitutes but as collaborators, enabling them to reimagine workflows, experiment with new ideas, and push artistic and intellectual boundaries.

By embedding their values, cultural contexts, and unique perspectives into their work, creators ensure that technology amplifies their voice rather than dilutes it. This approach reaffirms creativity as both a deeply personal and a collective act that bridges the past, present, and future.

Navigating Thoughtfully and Ethically

Integrating AI into creative fields requires a thoughtful and ethical approach. Empowered creators must navigate this landscape with awareness, addressing issues such as algorithmic bias, inclusivity, and accessibility. Institutions, industries, and creators share the responsibility of fostering an environment where innovation thrives without compromising ethical integrity.

By adhering to ethical practices, creators ensure their work aligns with values such as diversity, sustainability, and fairness. This foundation of ethical creativity not only enhances individual projects but also sets a standard for the broader creative ecosystem, inspiring trust and respect.

Redefining Boundaries

The machine age is not the end of traditional creativity but the beginning of a new chapter. It provides creators with tools to explore uncharted territories, break conventional barriers, and redefine what is possible in their craft. Whether blending AI-driven techniques with traditional methods or leveraging technology to solve complex problems, empowered creators are poised to lead a transformative era of innovation.

A Platform for Empowerment

Empowering creators in the machine age extends beyond technological proficiency. It is about cultivating resilience, adaptability, and a commitment to meaningful expression. By fostering a culture of collaboration and embracing the opportunities of AI while preserving the timeless essence of human creativity, creators can shape a future where their work inspires, connects, and elevates humanity.

The machine age is not just a challenge to navigate but a platform for empowerment. It provides creators with the tools and opportunities to expand their horizons, redefine their craft, and contribute to a more inspired, innovative, and connected world.

Summary

This chapter explores how creators can harness the opportunities presented by the machine age while maintaining the authenticity and emotional depth that define human creativity. It emphasizes the importance of AI literacy, balanced creative identity, and ethical innovation to navigate a rapidly evolving technological landscape. By leveraging collaboration between humans and machines, creators can unlock unprecedented potential while safeguarding the essence of their work.

Key Themes:

Building AI Literacy:

Foundational Understanding: Creators must understand key concepts of AI, such as machine learning and generative algorithms, to use these tools effectively. A basic grasp of how AI works empowers creators to integrate it into their processes meaningfully.

Critical Thinking and Ethics: Creators should remain vigilant about biases, fairness, and inclusivity in AI outputs, ensuring that their work aligns with ethical principles. Transparency about AI's role in the creative process builds trust and authenticity.

Developing a Balanced Creative Identity:

Preserving Human Authenticity: Creativity must remain rooted in human experiences, emotions, and cultural contexts. AI should complement, not replace, the creator's unique voice.

Combining Strengths: The synergy between human imagination and machine precision enables breakthroughs. For instance, AI can generate patterns or melodies, but humans shape these outputs into meaningful works.

Adaptability and Growth: Embracing a growth mindset allows creators to experiment with new technologies and reimagine workflows. Continuous learning ensures relevance in a tech-driven world.

Balancing Innovation with Ethics: Creators must address ethical concerns like intellectual property and bias while leveraging technology. This balance ensures innovation aligns with societal values.

A Future of Empowered Creators:

Collaboration for Innovation: The partnership between humans and machines fosters creativity that transcends traditional limitations. AI supports creators in tasks like data analysis and pattern generation, allowing them to focus on higher-order creativity.

Preserving Creativity's Essence: Technology must amplify human intent rather than overshadow it. By embedding personal and cultural perspectives into their work, creators ensure authenticity remains central.

Navigating Thoughtfully and Ethically: Empowered creators consider the broader implications of their innovations, promoting inclusivity and societal well-being.

Redefining Boundaries: The machine age offers tools to explore new creative frontiers. By blending traditional methods with emerging technologies, creators can redefine their craft and contribute to meaningful change.

Thriving in a Tech-Driven World:

Lifelong Learning: Creators must stay curious and informed about emerging tools while seeking interdisciplinary knowledge to fuel innovation.

Collaborative Networks: Engaging with diverse communities and global platforms broadens creative horizons and fosters cross-disciplinary partnerships.

Balancing Tradition and Innovation: Integrating technology thoughtfully with cultural heritage ensures that creative work resonates deeply with audiences.

Ethical Creativity: Responsible use of technology safeguards creative integrity and contributes to a fair and inclusive society.

This Chapter paints a vision of empowered creators who leverage AI to enhance, rather than compromise, their creativity. By cultivating AI literacy, embracing collaboration, and navigating the machine age with ethical intent, creators can redefine the boundaries of their craft. This approach positions creativity not only as a means of self-expression but also as a force for progress, connection, and societal transformation in an interconnected world.

13: The Universal Skill of Creativity

Creativity is often viewed as an innate talent, but in reality, it is a universal skill that can be cultivated and refined. In the context of a tech-driven world, where rapid change is the norm, creativity becomes even more essential—not just as a means of artistic expression but as a way to adapt, solve problems, and thrive in evolving landscapes. By fostering adaptability, critical thinking, and a mindset attuned to lifelong learning, individuals can harness the universal skill of creativity to navigate and shape the future.

13.1 Cultivating Adaptability and Critical Thinking

Adaptability and critical thinking form the cornerstone of creativity, enabling individuals to respond to challenges with resilience, innovate amid uncertainty, and continually learn from experience. In a technology-driven world, where creativity often intersects with rapidly evolving tools, these qualities are essential for addressing complex problems and seizing new opportunities.

The Role of Adaptability in Creative Growth

Adaptability empowers individuals to embrace change, experiment with new ideas, and learn from setbacks. It transforms uncertainty into a fertile ground for innovation by fostering an open and exploratory mindset.

For example:

✓ **In the Workplace:** Adaptability allows employees to integrate AI tools into their workflows, uncovering more efficient ways to achieve their goals and enhance productivity.

✓ **In Education:** Students who adapt quickly to digital platforms can fully leverage personalized learning technologies, improving both their understanding and academic performance.

Cultivating adaptability involves:

✓ **Staying Curious:** Approach new technologies and methodologies with curiosity, viewing them as opportunities rather than obstacles.

✓ **Practicing Flexibility:** Experiment with unfamiliar tools, techniques, or perspectives, even if they fall outside your comfort zone.

✓ **Embracing Change:** Recognize that growth often arises from moments of disruption or uncertainty, turning challenges into opportunities for development.

By fostering adaptability, individuals can remain agile and innovative, ready to navigate the dynamic intersections of creativity and technology.

Critical Thinking: A Tool for Navigating Complexity

Critical thinking complements adaptability by promoting a thoughtful and analytical approach to decision-making and problem-solving. It involves questioning assumptions, evaluating evidence, and considering the broader implications of actions, ensuring that creativity is purposeful and aligned with ethical principles.

For instance:

✓ **In Business:** Critical thinking enables leaders to assess the ethical implications of deploying AI systems, ensuring innovations are both impactful and aligned with societal values.

✓ **In the Arts:** Creators can use critical thinking to push boundaries while preserving authenticity and emotional depth in their work.

Enhancing critical thinking requires:

✓ **Asking Questions:** Challenge the status quo, explore alternative viewpoints, and remain curious about underlying assumptions.

✓ **Evaluating Sources:** Develop the ability to distinguish credible information from biased or unreliable content, particularly when working with AI-generated material.

✓ **Thinking Long-Term:** Consider the broader impact of creative decisions on society, culture, and the environment, ensuring sustainable and meaningful outcomes.

By combining adaptability and critical thinking, individuals can confidently navigate the complexities of a tech-driven world, creating work that is both innovative and ethically grounded. This balanced approach equips them to thrive in an era where creativity must constantly evolve to meet emerging challenges and opportunities.

13.2 A Guide to Thriving in a Tech-Driven World

Creativity in the machine age extends beyond artistic expression—it has become a crucial survival skill. It empowers individuals to adapt, innovate, and thrive in an interconnected, rapidly changing world. Adopting a proactive approach to learning and collaboration positions creators as leaders, ready to tackle challenges and seize new opportunities in this evolving landscape.

Embrace Lifelong Learning

The relentless pace of technological advancements demands a commitment to continuous learning. Lifelong learning not only keeps individuals competitive but also nurtures personal growth and ensures that creativity remains vibrant and purposeful.

Explore New Technologies: Staying informed about emerging tools and platforms is essential. Experimenting with AI-driven design software or augmented reality applications can open new creative possibilities. For instance, a filmmaker incorporating virtual reality into storytelling can create immersive experiences for audiences.

Seek Interdisciplinary Knowledge: Innovation often emerges at the crossroads of disciplines. Combining insights from diverse fields—such as engineering with the arts or technology with psychology—can lead to groundbreaking ideas that challenge conventional approaches.

Engage in Self-Reflection: Regularly assess your strengths and identify areas for improvement. Setting clear goals, such as mastering a new skill or revisiting foundational concepts, keeps your growth intentional and dynamic.

Through lifelong learning, creators can remain agile and ready to adapt their skills to the demands of a tech-driven world.

Build Collaborative Networks

Collaboration is the lifeblood of creativity. Engaging with diverse individuals and communities provides fresh perspectives, shared resources, and opportunities to co-develop innovative solutions.

Join Creative Communities: Participating in workshops, hackathons, and online forums fosters collective innovation. Collaborations often lead to unexpected synergies, such as a visual artist working with a data scientist to create compelling data visualizations.

Foster Mentorships: Building relationships with experienced mentors offers invaluable guidance, while mentoring others nurtures mutual growth. This exchange of knowledge reinforces the collaborative spirit necessary for success in a tech-driven world.

Leverage Global Platforms: Digital tools make it easier to connect with peers and audiences worldwide. Platforms like social media and international competitions expand reach, enabling creators to engage with diverse ideas and perspectives.

Collaborative networks amplify creative potential, transforming individual efforts into impactful collective achievements.

Balance Tradition with Innovation

While technology provides unprecedented tools, traditional methods and values remain essential anchors. Blending these elements ensures that creative work is innovative yet deeply authentic.

Integrate Tools Thoughtfully: Technology should enhance, not overshadow, creativity. Using tools like AI as extensions of the creative process keeps the human voice central. For instance, an author might use AI to draft character outlines while shaping the narrative with their storytelling style.

Preserve Cultural Identity: Creativity resonates most when it reflects personal and cultural contexts. Infusing work with heritage, traditions, or personal history adds authenticity and depth.

Experiment Boldly: Combining traditional techniques with modern technologies can yield groundbreaking results. A sculptor, for example, might use 3D printing to create intricate designs while maintaining the tactile artistry of hand-finishing.

By harmonizing tradition with innovation, creators can produce work that is both cutting-edge and rooted in meaning.

Prioritize Ethical Creativity

In a tech-driven world, creativity must be guided by ethical principles. Responsible use of technology preserves the integrity of creative work and contributes to a fairer, more inclusive society.

Champion Inclusivity: Creativity should reflect diverse perspectives and address underserved communities. For example, an app developer might design tools accessible to users with disabilities, ensuring equitable innovation.

Acknowledge Influences: Being transparent about the role of technology and external inspirations fosters trust. Crediting AI tools or cultural motifs openly adds value and authenticity to creative work.

Contribute to Societal Good: Creativity is a powerful force for addressing global challenges. Using skills to tackle issues like climate change or social justice can produce work that inspires change and drives tangible impact.

Ethical creativity ensures that innovation uplifts and benefits society as a whole, creating a legacy of positive change.

A Future of Creative Empowerment

To thrive in the machine age, individuals must combine adaptability, collaboration, and ethical awareness. By embracing lifelong learning, building collaborative networks, balancing tradition with innovation, and prioritizing ethical creativity, creators can unlock their full potential and make meaningful contributions to society.

In this evolving landscape, creativity transcends personal expression to become a universal skill—a driver of progress, a catalyst for connection, and a tool for shaping a more inspired and equitable future.

13.3 Creativity as a Universal Skill

Creativity is more than a talent reserved for artists or innovators; it is a universal skill inherent to all humans. It transcends boundaries, disciplines, and technologies, acting as a catalyst for self-expression, innovation, and connection. Whether solving complex problems, designing new products, or writing a novel, creativity plays a vital role in every facet of human life. In the machine age, where technological advancements are reshaping the way we live and work, creativity is not just a valuable asset—it is a necessity.

Creativity serves as a bridge between the familiar and the unknown, enabling individuals to navigate uncertainty and envision possibilities that do not yet exist. It empowers people to challenge conventions, explore diverse perspectives, and craft solutions that are as meaningful as they are innovative. This skill is not confined to any one profession or pursuit; it thrives in classrooms, boardrooms, laboratories, and studios alike. Its universality makes it a defining characteristic of humanity, uniting us across cultures, industries, and eras.

A Tool for Self-Expression and Innovation

At its core, creativity is a deeply personal act of self-expression. It allows individuals to channel their emotions, experiences, and ideas into tangible forms, whether through art, writing, technology, or problem-solving. This act of creation not only brings joy and fulfillment but also serves as a way to communicate with others on a deeper level. A poem, a painting, or a groundbreaking invention can resonate across time and space, connecting people through shared experiences and aspirations.

Creativity also drives innovation. It is the spark behind every transformative idea and groundbreaking discovery. From developing sustainable energy solutions to designing AI systems that enhance human lives, creativity fuels progress. In a world characterized by rapid change, the ability to think creatively ensures that individuals and organizations can adapt, innovate, and thrive.

Thriving in the Machine Age

The machine age presents both immense opportunities and significant challenges. AI and other technologies are expanding the boundaries of what is possible, but they also require humans to adapt and evolve. In this context, creativity becomes a survival skill—a way to navigate the complexities of a tech-driven world while preserving the essence of what makes us human.

To thrive, individuals must cultivate adaptability, critical thinking, and a commitment to lifelong learning. These qualities enable them to harness the potential of technology without being overshadowed by it. For instance, adaptability allows people to embrace new tools and workflows, while critical thinking ensures that their creative endeavors are purposeful and ethical. Lifelong learning fosters continuous growth, equipping individuals to stay ahead of trends and contribute meaningfully to their fields.

Rooted in Humanity

No matter how advanced technology becomes, creativity will always be uniquely and profoundly human. Machines can mimic patterns, analyze data, and even generate original-seeming content, but they lack the emotional depth, intuition, and context that define human creativity. It is this human element—the ability to infuse work with meaning, values, and intention—that ensures creativity remains a vital force in shaping the future.

By recognizing creativity as a universal skill, we can empower individuals to realize their potential and make meaningful contributions to society. It is a skill that transcends age, profession, and culture, offering a way to navigate the challenges of the present and build a brighter, more inclusive future.

Creativity: A Force for Good

Beyond personal and professional fulfillment, creativity has the power to address global challenges and inspire collective progress. From tackling climate change to promoting social justice, it enables individuals and communities to envision and create a better world. This transformative potential underscores why creativity must be nurtured and celebrated as a universal skill.

As we look to the future, creativity will continue to be the heartbeat of humanity—a force that inspires innovation, fosters connection, and shapes a world where technology and human ingenuity coexist harmoniously. In this ever-changing landscape, the universal skill of creativity remains not only a gift but also a responsibility, ensuring that progress is both meaningful and inclusive.

Conclusion

"Every masterpiece, whether carved in stone or coded in silicon, is a testament to the timeless resilience of human creativity."

14: The Evolving Spirit of Creativity

Creativity has always been humanity's most profound and enduring gift, evolving alongside us as we adapt to new eras, technologies, and challenges. It is a living force, continuously reshaped by the tools we create, the problems we face, and the dreams we pursue. In the machine age, where technology has become both a collaborator and a disruptor, the spirit of creativity is undergoing a transformation. Yet, at its core, it remains an unchanging essence—an expression of our humanity, our aspirations, and our desire to make sense of the world around us.

14.1 Embracing Change without Losing Essence

The pace of change in the modern world is staggering, driven by advancements in AI, automation, and unprecedented levels of connectivity. What once seemed like science fiction—AI composing symphonies, crafting paintings, or writing novels—is now a part of our reality. These technologies are reshaping how we think, create, and interact, offering tools that expand the boundaries of possibility. Automated storytelling, generative art, and algorithmic problem-solving are no longer confined to theoretical discussions; they have become everyday tools, challenging our traditional understanding of creativity, originality, and authorship.

Yet, amid this technological evolution, it is crucial to hold on to the essence of creativity. Creativity has always been more than the sum of the tools used to bring it to life. It is deeply rooted in human emotion, intent, and meaning. A painting is not merely about the strokes on a canvas but the artist's story and emotion that infuse it with life. A melody that stirs the soul does so because of the humanity behind its creation—the experiences, struggles, and aspirations that shaped it. Creativity is, and always will be, a reflection of the human spirit, not the tools employed to produce it.

To navigate this new landscape, creators must embrace change with openness and intention. This means seeing technology not as a replacement for human ingenuity but as an enabler—a tool that amplifies what we can achieve. AI might suggest new design iterations or help generate ideas, but the heart of the creative act remains human. It is the creator's personal vision, intuition, and emotion that transform an output into something meaningful.

Preserving traditions, values, and cultural contexts is equally important. While technology offers groundbreaking opportunities, it should enhance, not overshadow, the depth and authenticity that come from cultural and personal narratives. For example, a filmmaker using AI to refine a script must ensure the final work resonates with human experiences, emotions, and the societal values it seeks to represent. Similarly, an architect using algorithms to draft designs must infuse them with the human understanding of space, culture, and history that makes a building more than just a structure.

Creativity thrives at the intersection of balance—between tradition and innovation, intuition and precision, the individual and the collective. By maintaining this equilibrium, we can unlock the transformative power of technology while preserving the essence of what makes creativity uniquely human. It is in this balance that creators find not only inspiration but also a sense of purpose, ensuring that their work resonates deeply in a world where change is the only constant.

As the tools of creation evolve, the challenge is not to resist them but to integrate them thoughtfully. This integration ensures that the creative process remains a deeply personal and meaningful journey, where technology acts as a partner rather than a replacement. In doing so, we can embrace the opportunities of the machine age without losing the essence of creativity—the heartbeat of our shared humanity.

14.2 A Hopeful Vision for the Future of Creativity

EXTENSION: FROM BRUSH TO ALGORITHM

The future of creativity holds immense promise, fueled by the collaboration between human ingenuity and machine intelligence. This partnership is set to redefine the boundaries of what is possible, blending the precision and efficiency of algorithms with the limitless imagination of the human mind. Together, humans and machines can create not only art and stories but also solutions to some of the most pressing challenges of our time. This vision is one of inclusivity, accessibility, and transformative potential—a world where creativity knows no limits and connects us in ways we've yet to fully imagine.

Picture a world where a young artist in a remote village uses AI to animate their vivid imagination, crafting visual masterpieces that reach global audiences despite limited resources. In this world, a scientist and a poet collaborate to create works that fuse hard data with profound emotion, making complex knowledge more engaging and understandable for people from all walks of life. These are not distant dreams but tangible possibilities in a future where creativity transcends geographic, cultural, and economic barriers, giving everyone the opportunity to contribute their voice to a global dialogue.

This hopeful vision of creativity is not just about individual expression; it is also about collective progress. AI can democratize access to creative tools, making technologies once reserved for elite institutions available to anyone with curiosity and a desire to create. A filmmaker in a developing country could use AI-driven editing software to craft a compelling documentary. An educator could design immersive, interactive learning experiences with AI, reaching students in ways previously unimaginable. By opening doors that were once closed, this future has the power to amplify voices that have long been silenced or ignored, fostering a more inclusive and diverse creative ecosystem.

At its heart, this vision sees creativity not just as an end but as a means to solve global challenges. Artists and technologists could collaborate to raise awareness about climate change, using generative tools to visualize its impact in compelling ways. Musicians and AI systems might work together to compose symphonies that unite people across cultures, inspiring collective action for social justice or environmental preservation. The potential for creativity to drive meaningful progress has never been greater, and its applications stretch far beyond traditional art forms.

However, with great potential comes great responsibility. As we embrace these tools and the opportunities they bring, we must remain vigilant about the ethical implications of their use. Issues such as algorithmic bias, intellectual property, and the potential for homogenization must be addressed thoughtfully. The goal should always be to amplify human creativity, not dilute or replace it. We must ensure that inclusivity remains a cornerstone of this vision, creating a future where technology serves as a bridge rather than a barrier.

Preserving human dignity and authenticity is paramount. Creativity has always been a force for empathy and understanding, allowing us to see the world through the eyes of another. As we integrate AI into our creative processes, this human element must remain at the core, ensuring that the works we produce continue to inspire connection, reflection, and growth.

This hopeful vision for the future of creativity is one of collaboration, empowerment, and shared progress. It invites us to imagine not just a better way of creating but a better way of living—a world where creativity becomes a unifying force that celebrates our shared humanity while embracing the possibilities of the machine age. By navigating this future thoughtfully and ethically, we can ensure that creativity continues to be a beacon of hope, innovation, and meaning for generations to come.

14.3 Creativity: The Eternal Flame

Creativity is an eternal flame, a force that transcends time, space, and generations, illuminating the path of progress and human connection. It is the thread that weaves through history, linking the cave paintings of our ancestors to the digital masterpieces of today. In its essence, creativity is a reflection of who we are—a mirror of our times—and a guide toward who we have the potential to become. It bridges the past and the future, capturing the struggles and triumphs of humanity while envisioning what lies beyond the horizon.

This eternal flame burns brightest when it adapts to the changing winds of time. Creativity evolves alongside humanity, embracing new tools, perspectives, and challenges. In the modern age, AI and technological innovation are shaping the contours of creativity, offering unprecedented opportunities to explore, experiment, and express. Yet, even as we adopt these advancements, the core of creativity remains unchanged. It is rooted in emotion, intuition, and the deeply human desire to connect, communicate, and contribute.

The evolving spirit of creativity is an invitation—an open call to reimagine what is possible. It encourages us to break free from limitations, challenge conventions, and venture into uncharted territories. But this journey is not about abandoning the traditions, values, and cultural richness that define us. Instead, it is about carrying them forward, enriched and empowered by the tools and opportunities of a new era. Creativity thrives at this intersection of tradition and innovation, blending the timeless with the cutting-edge to create something uniquely powerful.

Creativity is also a force for unity and progress. It is the language that transcends borders, the spark that ignites movements, and the catalyst for solutions to humanity's greatest challenges. Whether through art, science, or technology, creativity connects us, reminding us of our shared humanity. In its most transformative form, creativity is a force for good, inspiring empathy, understanding, and collective action.

As we stand on the cusp of a creative revolution, driven by AI and other technological advancements, one truth remains clear: the essence of creativity is timeless. It is the heartbeat of humanity, a testament to our resilience, ingenuity, and boundless imagination. It is the song we sing, the stories we tell, and the dreams we dare to pursue.

To honor this eternal flame, we must approach the future with intention and care. We must embrace the tools of the machine age without losing sight of the human spirit that fuels creativity. This means preserving authenticity, upholding ethical principles, and ensuring that creativity remains inclusive and accessible to all. By doing so, we ensure that the flame continues to burn brightly, lighting the way for generations to come.

Creativity is not just a skill or a process—it is a legacy. It is the gift we inherit and the gift we pass on, shaping a future that is as hopeful as it is extraordinary. By honoring its spirit and nurturing its growth, we can ensure that creativity continues to inspire, connect, and transform, building a world where the possibilities are as infinite as our imagination.

Appendices

The appendices provide additional resources and practical tools to deepen your understanding of the ideas explored in this book. From real-world examples of human-AI collaboration to curated reading lists and actionable exercises, these materials are designed to inspire and empower your creative journey in the machine age.

Case Studies: Success Stories of Human-AI Collaboration

1. **Healthcare Innovation:** AI-powered diagnostic tools have revolutionized patient care. For example, DeepMind's AI-assisted detection of eye diseases allowed doctors at Moorfields Eye Hospital in London to identify complex conditions earlier and with greater accuracy. While the AI analyzed thousands of medical images, human specialists ensured the diagnosis aligned with patient needs and contexts, exemplifying how collaboration between humans and machines leads to life-saving advancements.
2. **Creative Arts:** Award-winning composer Holly Herndon collaborated with an AI program she co-created, affectionately named "Spawn," to produce her album *PROTO.* The AI-generated sounds acted as an extension of her artistic expression, blending the precision of machine-generated tones with her personal creativity, resulting in a truly unique musical experience.
3. **Environmental Solutions:** The collaboration between AI and conservationists has significantly improved efforts to combat climate change. In a project led by IBM, AI analyzed satellite data to track deforestation patterns, enabling human experts to target

conservation efforts with greater precision. This partnership showcased how AI can amplify the scale and impact of human-led environmental initiatives.

These success stories highlight the transformative potential of human-AI collaboration, demonstrating how blending human creativity with machine precision can address complex challenges and open new possibilities.

Recommended Reading and Resources on Creativity and AI

Expand your understanding of creativity in the machine age with this curated list of books, articles, and tools:

Books:

The Creativity Code by Marcus du Sautoy: An exploration of how AI is reshaping creativity in art, music, and beyond.

You Look Like a Thing and I Love You by Janelle Shane: A humorous and insightful look into the quirks and potential of AI.

Deep Thinking by Garry Kasparov: The chess grandmaster's reflections on his historic match against AI and the future of human-machine collaboration.

Articles:

"AI and the Future of Creativity" – *Harvard Business Review*: A comprehensive analysis of how AI is influencing creative industries.

"Can Machines Be Truly Creative?" – *The Atlantic*: A thought-provoking discussion on originality and machine-generated art.

Online Resources:

Coursera's *AI for Everyone* course: An accessible introduction to the basics of AI for non-technical audiences.

Kaggle's open datasets: A platform for exploring AI projects and datasets across various domains.

Runway ML: A user-friendly tool for experimenting with AI in art, design, and storytelling.

These resources provide a foundation for understanding AI's role in creativity while inspiring you to explore its possibilities further.

Practical Worksheets: Mapping Your Creative Process

These worksheets are designed to help you reflect on your creative practices and explore ways to integrate new tools and approaches effectively.

1. **Identifying Your Strengths:**

✓ List three aspects of your creative process that you feel are your strongest.

✓ Reflect on how these strengths can complement AI tools or collaborations.

1. **Exploring New Tools:**

✓ Identify one AI tool relevant to your field (e.g., DALL-E for visual arts, ChatGPT for writing).

✓ Experiment with the tool and note how it enhances or challenges your creative process.

1. **Balancing Tradition and Innovation:**

✓ Describe a traditional method or technique central to your work.

✓ Brainstorm ways to integrate technology into this method without compromising its essence.

1. **Setting Creative Goals:**

✓ Define a specific project or goal you want to achieve using human-AI collaboration.

✓ Outline the steps needed to balance your unique vision with the capabilities of AI.

1. **Ethical Creativity Checklist:**

✓ Before starting a project, ask yourself:

- Is the AI tool you're using transparent and explainable?
- Does your work promote inclusivity and fairness?
- Are you maintaining your authentic voice throughout the process?

By using these worksheets, you can refine your approach to creativity in the machine age, ensuring that your work remains innovative, ethical, and true to your personal vision.

The appendices serve as both a practical guide and a source of inspiration, equipping you with the knowledge, tools, and confidence to thrive as a creator in an ever-evolving world.

Acknowledgment

This book is the culmination of countless conversations, collaborations, and moments of inspiration that have shaped my understanding of creativity in the machine age. It is a reflection of not just my own thoughts, but the collective wisdom, support, and encouragement of many individuals and communities who have contributed to this journey.

First and foremost, I extend my deepest gratitude to my family for their unwavering belief in my vision. Your patience, love, and encouragement have been my guiding light, reminding me of the importance of perseverance and purpose.

To my mentors and peers, who have constantly challenged my thinking and broadened my horizons, thank you for your invaluable insights and thought-provoking discussions. Your diverse perspectives have enriched this book and deepened my understanding of the transformative power of human creativity.

I am especially grateful to the countless creators, technologists, and educators whose innovative work continues to inspire me. Your fearless exploration of the intersection between humanity and technology serves as a testament to the limitless potential of collaboration.

To the readers, thank you for embarking on this journey with me. Your curiosity and engagement drive the dialogue that keeps creativity alive and evolving. My hope is that this book sparks your imagination, challenges your assumptions, and empowers you to embrace the opportunities of the machine age.

Lastly, a heartfelt thanks to all the unseen hands—editors, designers, and technical experts—who have helped bring this book to life. Your dedication and expertise are the quiet forces behind every page.

This acknowledgment is not just a note of gratitude but a celebration of the collective spirit that fuels creativity. Together, we are shaping a world where imagination and innovation thrive, bridging the gap between what is and what can be.

Thank you all for being part of this journey.

With gratitude,

Sandeep Chavan

About the Author

Sandeep Chavan is a visionary thinker, educator, and author with a rich blend of experience in engineering, education, and writing. A mechanical engineering graduate with over two decades of experience in the industry and academia, Sandeep has dedicated his life to exploring the intersection of technology, humanity, and creativity. His multidisciplinary background lends a unique perspective to his work, enabling him to address complex topics with clarity, depth, and insight.

As an educator, Sandeep has spent years inspiring students in Science and Engineering, nurturing their potential and guiding them toward bright futures in engineering and other scientific domains. His teaching philosophy emphasizes critical thinking, adaptability, and lifelong learning—skills that are increasingly vital in the modern world.

Sandeep is also a prolific writer, known for his ability to distill complex ideas into accessible and engaging narratives. His works span diverse themes, including personal growth, professional development, geopolitics, globalization, business and economics and the evolving role of technology in human life. Through his writing, he seeks to empower individuals to navigate the challenges and opportunities of a rapidly changing world, with creativity and purpose at the forefront.

In this book, Sandeep combines his passion for creativity with his deep understanding of technology, offering a thoughtful exploration of how humanity can thrive in the machine age. His insights are rooted in his belief that creativity is not just a skill but a universal force that connects us all and drives progress.

When he is not writing or teaching, Sandeep enjoys exploring philosophy, engaging in thought-provoking discussions, and reflecting on the timeless wisdom of Indian heritage. His work continues to inspire readers and learners worldwide, encouraging them to embrace the future with curiosity, resilience, and an unwavering commitment to creativity.

For more information about Sandeep Chavan's work, visit: www.sandeepjchavan.com[1]

1. http://www.sandeepjchavan.com

Don't miss out!

Visit the website below and you can sign up to receive emails whenever SANDEEP CHAVAN publishes a new book. There's no charge and no obligation.

https://books2read.com/r/B-A-EPGPC-EXXKF

BOOKS 2 READ

Connecting independent readers to independent writers.

Did you love *Extension: From Brush to Algorithm*? Then you should read *Universal Skill*[2] by SANDEEP CHAVAN!

[3]

Universal Skill: Mastering the Art of Intelligent Action by Sandeep Chavan is a transformative exploration of a philosophy that is both timeless and urgently relevant in today's rapidly evolving world. The book delves into the concept of Universal Skill—a dynamic capability that transcends traditional notions of knowledge and habits, empowering individuals to navigate complexity, embrace change, and make thoughtful decisions across all aspects of life.

In an age marked by technological disruption, global interconnectedness, and constant change, Universal Skill emerges as the cornerstone for personal, professional, and societal success. Unlike static habits or specialized expertise, Universal Skill is a fluid and adaptive force, grounded in critical thinking, emotional intelligence, and ethical judgment.

---

2. https://books2read.com/u/4AlBAd

3. https://books2read.com/u/4AlBAd

Sandeep Chavan skillfully illustrates how this skill equips individuals to not only survive but thrive amid the uncertainties of modern life.

Drawing from history, contemporary examples, and cross-disciplinary insights, the book highlights real-world applications of Universal Skill. From Gandhi's non-violent revolution to the technological leadership of Sundar Pichai, Chavan provides vivid stories of great decisions that have shaped the course of humanity. These examples underscore how Universal Skill is not confined to any one domain—it is as crucial in shaping global policies as it is in building meaningful personal relationships.

The author offers practical tools, reflection prompts, and actionable strategies to help readers cultivate their own Universal Skill. He emphasizes its application in diverse settings: adapting to new professional roles, collaborating effectively with AI, resolving conflicts with wisdom, and fostering global harmony in a multicultural world.

What sets this book apart is its time relevance. In an era dominated by rapid technological shifts and cultural amalgamation, Universal Skill becomes the guiding philosophy for navigating unprecedented challenges and opportunities. Sandeep Chavan makes a compelling case for why mastering this skill is no longer optional—it is a necessity for leading a balanced, impactful, and purpose-driven life.

Whether you are a student, professional, leader, or lifelong learner, *Universal Skill: Mastering the Art of Intelligent Action* provides a roadmap for unlocking your potential and contributing meaningfully to the world. With its blend of thought-provoking insights and practical wisdom, this book inspires readers to embrace growth, act with intention, and create a legacy of positive change.

Read more at https://www.sandeepjchavan.com/.

Also by SANDEEP CHAVAN

Echoes of War: The Russia-Ukraine Conflict and Its Global Ripples
Shattered Lives
The War Economy

Ultra Modern Life Philosophy
The Duality of Laughter and Tears
Compare to Accept

Standalone
It's Not AI, It's AHI - Amplified Human Intelligence
The Decision Paradox
The Four Sapiens
Malicious Script of Indian Polity
The IIT Legacy & Global Impact
The IIT Legacy & Global Impact
You are just a Version of Your Original
Win the Game You Didn't Choose
Always Keep Your Bags Packed
Reveal to Shield - Learn the Game of Facades
Engage Beyond Elections
Culture, Identity & Change: The Evolution of Indian Society
The Triangular Dynamics

Never Trust Without Doubt
If You Can't Change the Country, Leave the Country
Live Fearless with Fear
India: The Investment Magnet
Realignment
Powering a Billion Dreams
Universal Skill
Extension: From Brush to Algorithm

Watch for more at https://www.sandeepjchavan.com/.

www.ingramcontent.com/pod-product-compliance
Lightning Source LLC
LaVergne TN
LVHW050301210926
848481LV00014BA/269